Big Bang Blasted!

The Story Of The Expanding Universe And How It Was Shown To Be Wrong.

Lyndon Ashmore

About the Author.

Lyndon Ashmore is a Physics teacher. He has an honours degree in Physics from the University of York (specialising in theoretical Physics) and an M. Phil research degree from what is now the University of Central Lancashire. His research project was in solid state Physics (solar cell technology in particular) and it was when he applied these laws of Physics to the Universe that he realised that observations normally associated with an expanding Universe could be better explained by the Physics of light interacting with matter. In particular this led to him showing that the important cosmological constant, the 'Hubble constant, H' is just a combination of three well known constants found on any schoolchild's calculator – Ashmore's Paradox.

The theory has been accepted for publication in the peer reviewed scientific journal "Galilean Electrodynamics.

Copyright © 2006 Lyndon Ashmore
All rights reserved.
ISBN 1 – 4196 – 3922 – 6

To order additional copies, please contact us.
BookSurge, LLC
www.booksurge.com
1 – 866 – 308 – 6235
orders@booksurge.com

In December 2004, an international team of scientists found that the larger of the 'hot' and 'cold' regions of the Cosmic Microwave Background (one of the 'proofs' of the Big Bang theory) were aligned with our own galactic plane. This means that these 'clumps' are not the seeds from which the galaxies formed but a local event instead. The Big Bang theory got it wrong!

In January 2005, a team of international scientists found a quasar whose redshift placed it at a distance of billions of light years away. The trouble is that it is embedded inside a galaxy that is known to be only 300 million light years away. The Big Bang theory got it wrong!

And then there is:

Experimental evidence tells us that the magnitude of the Hubble constant, H is related to the electron.

$$H = hr/m$$

per cubic metre of space

H = Hubble constant
h = Planck constant
r = classical radius of electron
m = rest mass of electron

This is nonsense - but this is what the experimental evidence is telling us. Therefore, it is the expanding universe theory that must be wrong.

Ashmore's paradox states:

Experiment tells us the Hubble constant is equal to the ratio of the radius of the electron to its mass, multiplied by the Planck constant, in each cubic metre of space. Therefore the Universe is not expanding.

Contents

Chapter -1.	Have you ever been lost?	Page 7
Chapter 0.	The Magic of the Spheres.	Page 11
Chapter 1.	How Far?	Page 17
Chapter 2.	Nobody Believes a Scientist!	Page 39
Chapter 3.	Is There Life on Mars?	Page 49
Chapter 4.	Blackpool Illuminations.	Page 61
Chapter 5.	Strange Stuff Is Light.	Page 83
Chapter 6.	Bar Codes and Bunsen Burners.	Page 103
Chapter 7.	Coffee in a Lay-by.	Page 113
Chapter 8.	Trumpets and Double Stars.	Page 127
Chapter 9.	Bangers and Mash.	Page 139
Chapter 10.	Pigeon Droppings and Nobel Prizes.	Page 163
Chapter 11.	Bring on the Inflating Crystal Spheres	Page 175
Chapter 12.	Now, This is What I Call a Big Bang!	Page 187
Chapter 13.	The Farmyard of Codsmology!	Page 195
Chapter 14.	Electron Rules OK?	Page 205
Chapter 15.	Bull in a China Shop.	Page 223
Chapter 16.	Tired Light and CMB.	Page 239
Chapter 17.	Supernovae Time Dilation.	Page 253
Chapter 18.	A Dialogue On The Two Chief World Systems	Page 267

Chapter negative 1. Have you ever been lost?

Have you ever been lost?
Have you ever looked out of your car window at some bleak and desolate spot and known that somewhere along the road, you must have taken the wrong turning?

You don't know where you went wrong but it is blatantly clear that you did, because a quick look outside confirms that this is not where you should have been.

Perhaps you were going to visit friends in their new home and had set out with instructions and map in hand.

You were not too sure about one turning some distance back but from then on, all went well.

There was the church on the right as shown on the map; then there was the garage exactly where it should have been, so you must have been right.

But you must have gone wrong somewhere, because the place where you have ended up is definitely not correct.

Instead of being at your friends' new home on a spanking new housing development you are in a deserted farmyard.

But you are sure that you followed the map.

However, it is obvious that you can't have, because this is just not where you should have been.

In the end you have to go right back to the beginning and retrace every step, checking and double checking every decision that you made along the way, until you find your mistake.

It is like that with the Big Bang theory and the expanding Universe.

We were a little unsure about the theory at first, but from then on it appeared that all the signs were in the right places. So we thought no, the expanding Universe theory must have been correct. The Universe started in a Big Bang and has been expanding from that point on. We became more and more certain of the theory as we went along but then we ended up somewhere where we definitely should not be! As in our car journey, it does not matter how sure we are about our route, we know we must have gone wrong somewhere because of where we ended up.

We have no choice but to go back to the beginning and retrace our steps to find out just where we went wrong.

On the car journey, it was when we finally ended up in the farmyard that finally convinced us, beyond any doubt, that we had gone wrong somewhere.

With the Big Bang theory, experiments now tell us that the magnitude of the Hubble constant can be expressed in terms of the electron and so we know that we have gone wrong somewhere along the route.

As we developed the Big Bang theory, technology has been constantly improving and we are able to measure the rate at which the Universe is supposedly expanding, the Hubble constant, to a greater and greater accuracy. It was then that I realized that the number we finally measured is, in fact, just our old friend the electron in disguise.

However, this can't be.

There is no way that the rate at which the Universe expands can be linked to the electron.

So the expanding Universe theory must be wrong!

In the Big Bang theory, the rate at which the Universe expands, the Hubble constant, is said to tell us the age of the Universe.

But experimental evidence tells us that the magnitude of the Hubble constant is just the electron in disguise and so if the expanding Universe theory is correct then the age of the Universe must be the electron in another guise.

But it can't be.

It is just nonsense.

No one will believe it.

But experiment is telling us that this is the case - there is no doubt about it!

So it must be the theory that is wrong. The Universe is not in fact, expanding.

Because experimental data now tells us that the magnitude of the Hubble constant is the electron in disguise, we are lost in the farmyard of cosmology and we have to go right back to the beginning and start all over again. Now we can say without doubt that the Big Bang theory and the expanding Universe is just a load of old CODSMOLOGY!

Welcome to Ashmore's paradox!

Chapter Zero. The magic of the spheres.

In retracing our route, we do not have to perform any of the experiments again. They are undisputedly correct.

It is the interpretation of these results that is in question. Somewhere along the route to the Big Bang theory, someone or some people have made a wrong assumption or attached the wrong interpretation to an experimental result. Consequently, it is the assumptions and interpretations of the experimental results that we have to re-examine at every step.

Theories are only good in that they agree with experimental results. It does not matter how good a theory makes you feel, or how beautiful the theory is or even whether the theory agrees with one's own religious beliefs, the fact of scientific life is that if the theory does not agree with experimental results then the theory is wrong and we must find another one.

That is why, with Ashmore's paradox now coming to light, we know that the Big Bang theory has to go.

The Hubble constant is supposed to be a kind of 'velocity', the rate at which far-flung galaxies are whizzing away from us. Out of this enormous initial explosion, the 'Big Bang', space itself is supposed to be expanding carrying with it the matter that eventually coagulated to the form planets, stars and galaxies. The rate of this supposed expansion can have any value at all and so it comes as quite a shock when we realize that it is related to our old friend and fundamental particle, the electron. It just can't be true, but no matter how many times one calculates and recalculates the values, the experimentally determined value of the Hubble constant is always equal to a combination of three common physical constants, in each cubic metre of space.

So we instantly know that the theory is wrong.

No matter how beautiful the Big Bang theory is, we know it has to go.

No matter how many famous scientists have stood up and defended it, now we see that the measured value of the Hubble constant is the same in every way to 'this much of an electron' in each cubic metre of space, then we know the theory is wrong. It has happened many times before.

Scientists have always been sure of themselves and have always thought that they knew what was going on. We now laugh at the ancient Greek's model of the Universe that had the Earth at the centre but, at that time, it explained perfectly the complicated motion of the heavens as they knew them. That is, the model agreed with the experimental results that they had at the time and so they believed the model to be correct.

In about 530 BC, Pythagoras collected together a group of intellectuals and they improved upon an earlier model of the Universe that had originally been put forward by Thales some seventy years earlier. Here, all the fixed stars were on a gigantic celestial or 'star' sphere, which rotated about the Earth, once every twenty-four hours. The Earth itself was a tiny sphere fixed at the centre of the big 'star sphere'. The Sun was fixed on a second crystal sphere, which carried the Sun through the heavens as the sphere turned. The Sun's crystal sphere was bolted in such a way to the fixed star sphere, that it too could rotate once every twenty-four hours around the Earth, but it could also move at an angle across the star pattern, in the same way that the Sun does in real life. The planets each had their own crystal sphere that allowed them to move across the heavens. The spheres had to be made of crystal so that they were transparent and thus one could see the fixed stars behind them. Unfortunately this did not match the observed motion of the stars and planets, so, in 370 BC, it had to be 'tweaked' by Eudoxus. The Sun was given three more crystal spheres bolted to each other in different places, so that the crystal sphere model reproduced the motion of the Sun exactly. Other crystal spheres were added so that the Moon and planets each had several crystal spheres of their own. The whole model of the Universe was like a giant onion, with one concentric crystal sphere inside another. In the end, there were fifty-five crystal spheres in all, but the point is, it described every astronomical observation known at the time. Aristotle himself believed in this theory and he also believed that the crystal spheres actually existed.

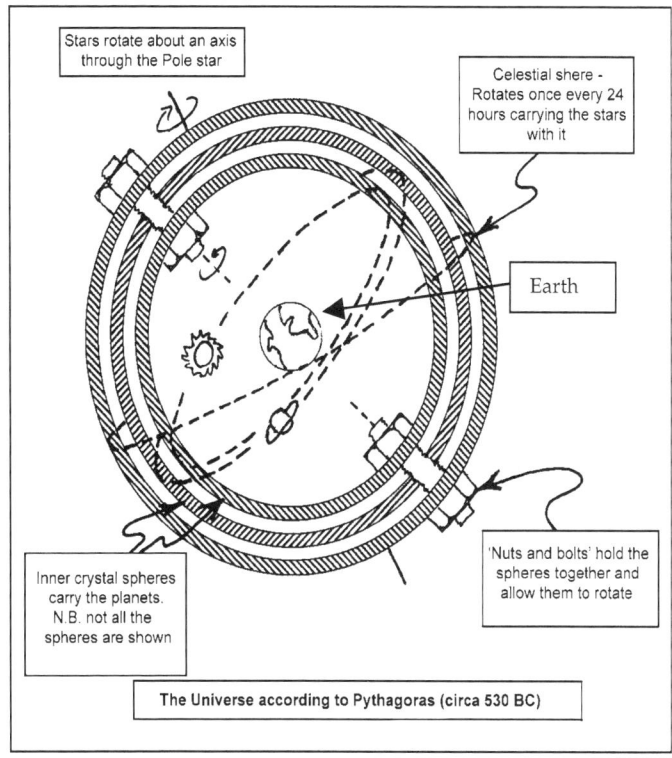

The Universe according to Pythagoras (circa 530 BC)

Scientists and philosophers would write a scroll on the theory; teachers would teach it as if it was correct - because as far as they were concerned, it was correct!

Imagine yourself in an ancient Greek school, with the teacher teaching astronomy to a group of students.

"Andreas, stop fiddling with your thobe and take a note of what I am saying".

"I can't take a note Sir; my clay tablet has all dried up. My scalpel won't make a mark on it".

"Then wet the clay, boy, wet the clay. The water is in the urn in the corner; that's what it's there for".

"Now, where were we? Ah yes! The Stars, Moon, Sun and planets all have their own transparent spheres and these spheres rotate around the Earth which is fixed in the centre".

"Please Sir".

"What now, Andreas?"

"What holds the transparent spheres together Sir?"

"Good question, boy. There are two nuts and bolts holding them together, one each above the North and South poles. This allows the spheres to rotate at different speeds whilst still giving us the daily rotation of the stars, planets etcetera, and etcetera about the Earth".

"Well Sir, why can't I see the nuts and bolts that hold the spheres together when I look up at the sky?"

"Because the Sun is too strong, boy. The Sun lights up the sky so much in the daytime, that the sky is too bright. The reason why you cannot see the Northern nut and bolt is because it is hidden behind the glorious blue sky given to us by our Gods up on Mount Olympus."

"But Sir, the Northern nut and bolt should be in the same place as the North Star. I can see the North star at night so why can't I see the Northern nut and bolt at night?"

"Because it's dark at night you silly boy, because it's dark! Look, pay attention you lot otherwise you will fail your examinations and end up on the end of an oar in one of Helen of Troy's ships".

"Must be better than teaching though, Sir!"

You see they thought they knew what was going on. In future years, people will look back at the period that is just ending and laugh.

"You mean they actually thought that all the matter in the Universe emerged from one tiny point in space? They must have been stupid"

But they will have more information and more experimental results than we have. It is easy to be clever in hindsight. On our car journey to the Hubble constant, the view through the rear view mirror is always clearer than the view through the front windscreen.

Scientists have always thought they knew what was going on. They always thought that the theory they had at the time was correct because it explained what they knew to be happening. It is only when something comes along which the theory clearly cannot explain, that scientists are forced to either tweak the old theory, or throw it out and find a new one.

It is the same with the Big Bang theory. Scientists believe it to be correct at this time. Museums have displays on it; television programmes are made about it, because up till now, they thought it was correct.

It had started off as a simple theory with an initial explosion, followed by an expansion of space but it has become more and more complicated as they tried to tweak the model to cope with new findings. 'Dark matter', 'negative energy' and 'cosmological constants' have all been brought in to try to save the sinking ship, in the same way that the Greeks had introduced more and more crystal spheres. However, now that observations tell us that the Hubble constant is just the electron in disguise, the theory has to be discarded because this relation is inconsistent with the Big Bang and the expanding Universe. No extra crystal spheres can save the Big Bang theory this time. According to the Big Bang theory, there is no way that the Hubble constant and the electron can be related. Some say that it could be 'by chance', a fluke of nature that these two quantities just happen to have the same values at this instant in time – but this is very unlikely.

With the Big Bang theory, the value of the Hubble constant changes as the Universe grows older. The effect of gravity is supposed to slow down the expansion and should cause the Hubble constant to decrease. 'Negative energy', another crystal sphere dreamt up by cosmologists, in an attempt to tweak the theory, to try to explain another unexpected experimental result, supposedly will cause the expansion to speed up and thus increase the Hubble constant.

Since the Hubble constant is supposed to be changing, it is argued that there must be a point in time when these two quantities, the Hubble constant and the parameters of the electron will be the same, just by chance.

The problem with this argument is that it places us at a special place in the history of the Universe and scientists have had their fingers burned too often when doing this in the past. It is like saying that by chance, the Earth is at the centre of the Universe. No way!

It is just too improbable that the first time we are able to measure the Hubble constant accurately, it just happens to be equal to a combination of the parameters of the electron.

Scientists worry greatly when two quantities are the same but have no relationship between them. Furthermore, now we know that it is the electron that is responsible for the Hubble constant and not the 'expansion of the Universe' then it is a relatively simple matter to work out the true relationship between them - and then there is no need for the 'Bangs' and 'expansions' of the Big Bang theory.

In the following pages we will retrace the route to the Hubble constant by starting at the very beginning. See if you can spot the place where the 'wrong turning' was made as we see how it really is, and how the electron is truly responsible for the Hubble constant.

This is a very serious piece of writing and everything you read here is true, both historically and scientifically.

Chapter 1. How far?

When I arrived at York railway station in 1968 and was bundled onto the back of an open lorry (provided free by the student's union it must be said) to be whisked away to start a degree course in Physics, I had little idea that over thirty years later, I would be writing a theory based, in part, on principles founded in this City. Arriving at Heslington Bog, the site of York University, one of the newer Universities at that time, frozen by the cold East wind that had teased us 'freshers' on the back of the lorry, trying to tempt us to turn around and go home (it wouldn't have needed much persuading actually), we were led to a desk where one or two people appeared to know what was happening.

"Who are you?" they asked with confidence.

"Ashmore" said I, meekly.

"Oh, you're in Goodricke College. It's just opened."

So off I trundled to find my room and was greeted by an invitation to 'attend the grand opening ceremony of Goodricke College.' I thought that I might as well attend this 'Grand opening' - mainly in the mistaken belief that there may be a few beers going free. The 'Grand Opening' consisted of Michael Swann FRS Vice chancellor of the University, the new Provost Prof. Woolfson - a Physicist, a handful of lecturers and a few students including myself, looking for the free beer.

"Who is John Goodricke?" whispered one of the new inmate students.

"Some astronomer guy", whispered back another "who used York Minster to take his Measurements"?

"Can you see through York Minster?" asked another.

The plaque was unveiled, a few words said and everyone drifted away wondering where the free beer was (probably with the provost and the lecturers!). So there I was, a member of a spanking new college named after one 'John Goodricke', never been used before and spotlessly clean (it was cleaner still that night as someone placed a packet of soap powder in the fountains causing them to lather up for the next week!), destined to spend the next three years studying Physics at the Grand old 'Uni' of York.

Why do I start my tale here? Well, it was this 'astronomer guy', Goodricke who actually started the trail to the Hubble constant, in York, some two hundred years earlier. One of the biggest problems in determining the Hubble constant is in finding out how far away heavenly bodies are.

For instance, how big is the Universe? How far away is the Sun? How about the Moon, how far away is that? People always knew it was a long way away, but just how far is 'a long way'? The very word 'astronomical' has the informal meaning, 'enormous amount' - but just how enormous is enormous?

The trouble is that a small object up close looks big whilst a big object far away looks small. If you look at something in the heavens and it looks small it just might be that it is small or it might be that it is actually very big but a long way away – you have no idea which one it is.

Lying on the sofa pretending to be a couch potato (but you aren't really because you are pondering the Universe), you look at the light bulb dangling from the ceiling. It looks very small. But, when you climb a ladder to change the bulb it grows bigger and bigger as you climb the ladder and get closer and closer to the lamp. This difference in apparent size is caused because the distance to the light bulb is similar to the height of the ladder. The nearer you are to something then the larger it appears to be. Look at the Moon and it looks small. Is it because it is small and close to you or is it really big, but a long way away?

Climb a ladder and it still looks the same size so it is much further away than the height of the ladder. Climb a mountain of a few thousand feet and the Moon still looks the same size – it doesn't appear to get any bigger at all. The distance to the Moon must be much, much greater than the height of the mountain. But how far away is it?

<u>Mathematics problem for the student</u>.

You are asked to find the width of a river. There are no bridges, boats or other means of crossing the river and, to make the problem more interesting, the river is infested with sharks and piranha fish so if you are clever enough to solve this problem you are clever enough not to dip your big toe (or any other appendage) into the water. However, you are provided with a tape measure and a protractor. How would you do it?

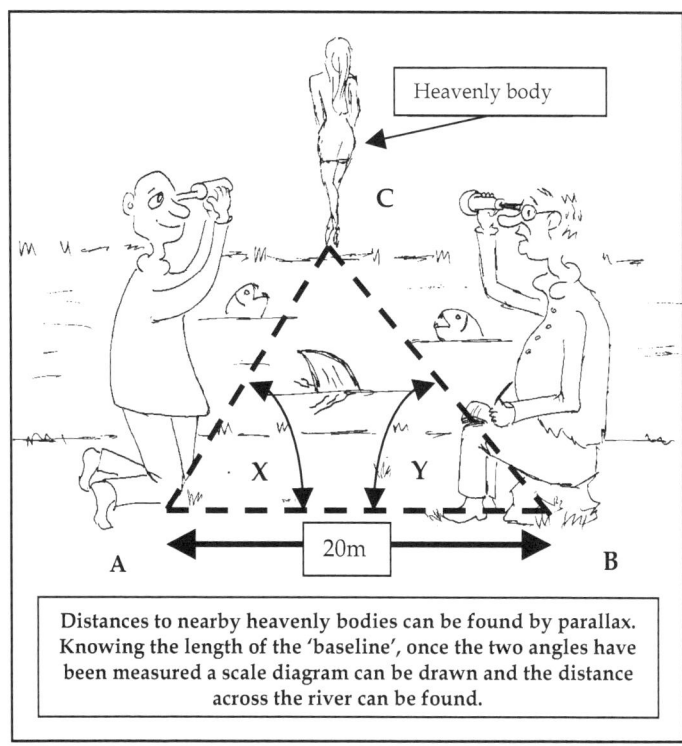

Distances to nearby heavenly bodies can be found by parallax. Knowing the length of the 'baseline', once the two angles have been measured a scale diagram can be drawn and the distance across the river can be found.

Easy, measure it by parallax of course. Measure out a baseline of about 20 metres on your side of the bank (which represents the Earth in the astronomical problem) and call it a 'baseline AB'. Pick a point, C on the other side, (this, in space represents our Heavenly body). Go to point A and measure the angle (x) a line between A and C makes with the riverbank. Go to point B and measure the angle (y) a line between B and C makes with the riverbank. You now have two angles and a baseline distance. Draw a scale diagram and determine the width of the river (or use trigonometry to calculate it if you like).

You can use this technique to measure the distance to the Moon. Because the Moon is a long way away, you need a bigger baseline (and remember you have to take both measurements at the same time as the Moon appears to move!). Hipparchus did this in about 150 BC using parallax measurements during a solar eclipse. He achieved a value only a few percent different than the one we use today. Finding the distances to the planets occurred much more recently as we had to wait for sensitive instruments to be devised and produced in order to measure the very small angles involved. In the late seventeenth century French astronomers just about managed to find the distance to Mars by parallax and gradually, with time, distances within the solar system were measured.

Having determined these distances, astronomers turned their attention to the stars. The trouble is that the stars are so far away that any baseline on Earth is too small to give measurable angles of parallax. But, the Earth revolves around the Sun and so we can use the diameter of the Earth's orbit as a baseline. As the Earth moves around the Sun the position of nearby stars appears to move across the distant star pattern.

In the following exercise, your nose represents the Sun, each eye represents the Earth at either end of a diameter of its orbit and the bottle and glass on the bar represents the background of distant stars – the celestial sphere.

Try holding up one finger at arms length so that it appears in front of the bottle.

Government Health Warning. Don't try this experiment in public as you could offend someone and seriously damage your health!

Keeping the other eye closed, look at your finger first with your right eye and then with your left eye. You will see that your finger appears to move across the background of desirable objects. This is parallax.

Now, bend your arm so that it is only half extended. Repeat the exercise, looking first with your right eye and then your left eye. You will see that your finger now appears to move further across the background objects. The closer the object or star is, then the greater the parallax or apparent movement against the background.

Hold one finger at arms length and view it first with one eye and then the other. You will see it appear to move across the backdrop of your room. Now only half extend your arm and repeat the exercise. Your finger will appear to 'move' much more. This is the effect of distance upon parallax.

Government Health Warning: This experiment could seriously damage your health – if done in public!

In this way, astronomers could measure parallax angles to nearby stars by taking two measurements, six months apart. Knowing the distance from the Earth to the Sun they could then determine how far away the star was.

This is fine for measuring the distances to the nearest stars but the problem is most stars, and especially galaxies, are so far away that even the diameter of the earth's orbit is too small to produce measurable parallax angles. Why not use the rotation of our galaxy, the Milky Way and use that as our baseline I hear you ask? Well, since the Milky Way galaxy rotates with a period of rotation of over 200 million years you would have to wait half that time between measurements - so I would not recommend it for a PhD thesis! No, as far as parallax is concerned, we are stuck when it comes to measuring the distances to stars or galaxies more than a few light years way.

This is where York and John Goodricke enter the picture. The Goodricke's were an aristocratic family with lots of money. John's father was a diplomat supposedly with distant links to royalty including William the first and Alfred the Great whilst his mum was the daughter of a Dutch merchant. John was born in Groningen in the Netherlands in 1764 but unfortunately he caught scarlet fever at an early age and that left him deaf and mute. Unfortunately, at this time anyone who was deaf was considered by society to be mentally incapacitated, or to be blunt, If you were deaf then no one usually bothered to educate you because they didn't think it worthwhile. Fortunately for John, his parents didn't believe in this and gave him a good education at a specialist school for the deaf and dumb in Edinburgh, Scotland – one of the first of its kind, where he received a basic education and learned how to lip-read. From Edinburgh he was enrolled at the 'Warrington Academy', a theological seminary where he studied under Joseph Priestly and William Enfield. Enfield was a tutor of religion and astronomy and Goodricke flourished in both Mathematics and Astronomy. After graduating from Warrington, he moved to York where he lived with his parents.

York was, and still is, a highly intellectual and scientific community and just a few doors away from the Goodrickes lived the Pigotts who just happened to own one of the best private observatories in England.

Nathaniel Pigott was an astronomer in his own right and his son; Edward followed his father by being a very keen amateur astronomer. The Pigotts had lived in France for a large part of their lives before moving to Glamorganshire and then on to York. Edward had already made a few discoveries of his own. Whilst in Glamorganshire, in 1781, he discovered a 'nebula' - though he was a little late in publishing the discovery and so he didn't really receive the credit he deserved. He had better luck in York where, in 1783, he not only discovered a comet but he was also given the credit for the discovery. Edward finally ended up dying in Bath (the City in England - not the tin bath with hot and cold running water used for washing).

So what with Goodricke's interest in astronomy and all this going on just down the road, it is little wonder that he took an interest in his neighbour's astronomical pursuits as here, in astronomy, his deafness was no handicap -he was as good as anyone else, if not better. Goodricke and Pigott set themselves the task of looking for variable stars, that is stars whose brightness varied over time.

People had known of the existence of variable stars for some time. David Fabricious was a pastor and, like many other pastors of his day he was an amateur scientist specialising in astronomy. He was born in Osteel, East Frisia (northwest Germany) and is credited with 'discovering' the first variable star 'Omicron Ceti' in 1596, which was later given the name 'Mira' (although there is evidence that the ancient Babylonians knew about it first).

Fabricious was looking for the Planet Mercury at the time but found Mira instead! (Mira, a name that means 'the wonderful' lies in the constellation of 'Cetus' the 'Sea Monster').

In case you are interested (it doesn't matter whether you are or you are not, I am going to tell you anyway) Greek mythology tells us that there was this King Cepheus and Queen Cassiopeia who had a daughter called Andromeda. All are names of constellations in the night sky. Cepheus is a faint constellation, unlike that of his wife, Cassiopeia, which is the bright 'W' near the Pole Star. This is as one would expect as Cepheus was a hen pecked husband and it was his wife, Cassiopeia, who was the big boss.

Cassiopeia was a gorgeous woman who would tell anyone who would listen to her what a cracker of a woman she was. She even started boasting that she was more beautiful than any of the women in King Neptune's underwater harem. Naturally, they were a bit upset about this so they cut off all of Neptune's favours and nagged him to do something to stop Cassiopeia insulting them. His response was to send a sea monster Cetus, to gobble up any women and children playing on the shores of Cassiopeia's kingdom.

Not surprisingly, the citizen's were not too happy about this, so King Cepheus went to see a soothsayer who told him that the only way to stop his populace being gobbled up was to feed his daughter Andromeda to the monster. Andromeda was laid on the rocky shore, chained by her ankles and wrists (I am saying nothing!) and left for Cetus to eat her all up. This is the position in which you see Andromeda in the night sky.

Cassiopeia, had to be punished too so she was fastened in her chair and placed upside down in the heavens, careering half a circle around the Pole Star every night. Being placed upside down in her chair was her sentence for her excessive vanity.

Coming back to reality, the nice thing about Mira is that its brightness changes so much that it can easily be seen by the naked eye when it is at its brightest but appears to completely disappear from the night sky when it goes dim. Fabricious thought it must have been an exploding star or 'novae' (meaning 'new'). It was only later that astronomers realized that Mira was a variable star, appearing suddenly every eleven months, remaining visible for around two months of its cycle and then disappearing again. Fabricious went on to record the first sunspots on the Sun but is alleged to have come to a sticky end over some fowl play. A crime wave broke out in his parish where a thief was going around stealing chickens and geese. Fabricious, in his role as pastor, delivered a sermon on the evil of this practice. Foolishly, he said he knew who the thief was and if the stealing did not stop he would do something about it.

Unfortunately for Fabricious, the crime wave escalated and he was found murdered the next day having been hit on the head with a shovel by the thief.

The discovery to the Western World of this variable star would have created a great scandal as this was before the time of Galileo and his inquisition. The church insisted that the Earth was at the centre of the Universe and everything else revolved around it. Furthermore, since God made everything in the Heavens, then every heavenly body found within it must also be perfect - for perfect, read 'spherical and without blemish'. To have a star that flashed on and off went against all the teachings of the church.

The role of variable stars in confirming the Copernicus theory that the Earth went around the Sun and not everything in the heavens was hunky dory is often forgotten. The first 'evidence' that Heavenly bodies were not perfect is usually credited to Galileo's telescopic observations of the mountains on the Moon but Fabricious' discovery of the 'non perfect' star Mira was well before this.

Jules Verne also mentions Fabricious in the book "From The Earth To The Moon". In the book, Verne cites Fabricious as having turned his telescope onto the Moon and seen little men on its surface. However, you will only find this in this book of fiction, as Jules Verne made up the story.

In 1667 an Italian astronomer Geminiano Montanari, became the first Western astronomer to notice that the star 'Algol' was changing in brightness. Algol is the second brightest star in the constellation Perseus and the name 'Algol' has the meaning 'Ghoul.' Geminiano was born in Modena, Italy in 1633 and was sent to Florence to study law as the family needed money and this was the only profession that paid good money in those days (nothing changes does it?). Unfortunately for Geminiano (or it could be said to be fortunate) he had an affair with a society lady and was thrown out of University. He later moved to Vienna where he received doctorates in Law, Philosophy and Medicine.

Geminiano dabbled in lots of different areas of science, he led a long campaign against astrology, drew his own a map of the Moon, studied hydraulics, wrote a manual for gunners, made lenses, successfully persuaded the Venetian state to divert all rivers away from the lagoon to stop it silting up and spent most of his later life looking after the mint. During all this he found time, in 1670, to write down that the brightness of the star Algol varied.

Algol is an Arabic name meaning 'Head of the Ghoul' and in fact, Algol is situated in the head of the Medusa in the constellation of Perseus.

Greek mythology tells us that in contrast to Cassiopeia, Medusa had snakes on her head instead of hair and was so revolting that anyone who looked at her was instantly turned to stone. Perseus was a knight in shining armour who went to kill Medusa. He had been given a special helmet, which made him invisible whenever he wore it and a magic pouch in which to hold Medusa's head once he had cut it off.

Invisible in his helmet and guided by the reflection in his shining shield, so as not to look directly at the horrible sight and be turned to stone, he cut off the head of Medusa and tucked the head into his magic pouch. Returning home triumphant on his trusty stead, the winged horse Pegasus, he noticed a ruckus on the beach below him. Zooming down for a closer look, he saw a beautiful young maiden, chained to the rocks on the beach and about to be eaten by a sea monster in front of a crowd of locals.

With impeccable timing, he shouted to Andromeda to close her eyes, he opened the pouch to give the sea monster Cephus a quick flash of Medusa's face. The sea monster was immediately turned to stone, Andromeda was freed, the crowd cheered and everyone lived happily ever after.

The reason for including these two episodes of Greek mythology into what is meant to be a serious book is that you will notice that in both cases, the variable stars had been included as the flashing eyes of 'monsters' within the constellations. Myra was placed in the sea monster, Cetus and Algol was the blinking eye of Medusa. This suggests that the ancients knew all along that these two stars were variable stars and that Fabricious and Montanari only rediscovered them for the Western World.

By 1715 a list of the six known variable stars was published by Edmond Halley. Pigott thought there must be more and so Pigott and Goodricke set out to find them. They became the first variable star astronomers and are sometimes known as 'the fathers of variable star astronomy'. They must have seemed a strange pair. Goodricke was 17 years old and, being totally deaf, he and Pigott would discuss things by passing notes (which is difficult in astronomy because it's night and too dark to read them). Pigott, older by about twelve years, insisted in wearing bright flowery clothes, a habit he brought with him from late 17th century France - so he must have looked out of place or a bit of a 'dandy' in York with its 'Northern' values and strong Quaker ties. Nevertheless, in just four short years, they made York into a world centre of astronomy.

Goodricke studied Montanari's star, Algol and calculated its period at 68 hours 50 minutes. Whilst Algol was not in the list published by Halley, Goodricke must have suspected that it was different to other stars since Algol was known in England as the 'Devil's Star'. What surprised Goodricke was how fast the brightness of Algol changed. He wrote that he watched it for an hour or so and he could see the change in brightness even over this short period. He found this so hard to believe that he first looked for other explanations such as an optical illusion or atmospheric effects or even something wrong with his eyes. However, the more he watched Algol the more he realised that it was the star itself that was changing in brightness.

Goodricke reported his findings to the British Royal Society, whilst Pigott kindly stood back to allow Goodricke to take the sole credit for the discovery - as they must have done the observations together.

Goodricke put forward two possible explanations for the variation in brightness. The first, and correct explanation as it turned out, was that Algol had a darker body revolving about it. The total brightness would be reduced whenever the dimmer of the two stars moved between Algol and the Earth. The newspapers of the time had a whale of a time with this and postulated that this meant that there were solar systems other than our own and speculated on the chances of alien life. It is interesting that this is now one of the ways that we actually do locate other solar systems around other stars – watching for their brightness to dim as a planet moves across our line of sight. However this is a much smaller effect than that found in binary stars like Algol. The second explanation proposed by Goodricke was that the star had one side or region darker than the other. As the star revolved, it would 'dim' whenever this darker side was facing to the Earth. This type of variable star is now also known to exist.

Pigott discovered the variability of Eta Aquilae (then known as Eta Antinoi – its name was later changed) and, in the same year, Goodricke discovered the variables Beta Lyrae (Sheliak) and Delta Cephei. Pigotts discovery of Eta Aquilae was the first known representative of a new type of variable star known as 'Delta Cephei stars' or just Cepheids (named after Goodricke's discovery of Delta Cepheid). If you are wondering why these stars have Greek letters as their first names then the answer lies in the Bayer system. In this system, the stars have as 'surname' the constellation in which they reside and their 'first name' depends upon how bright the star is. The stars in a constellation are placed in order of their brightness and named alpha, beta, gamma, delta and so on. Hence 'Delta Cephei' is the fourth brightest star in the constellation Cepheus.

Typical isn't it. You befriend the boy next-door, give him a hand and he ends up stealing all the glory! Why didn't they name them after Pigott's discovery, after all he was the first?

Who knows really, but it must be said that Pigott was unlucky again. His Variable star, Eta Aquila was called something else when he discovered it. It was known as 'Eta Antinoi' because it was in the constellation of Antinous and Antinous is now 'extinct'. Goodricke's variable star is in the constellation of king 'Cepheus' hence the names 'Delta Cephei' and 'Cepheid variable stars'. This constellation survives to this day. Also, Delta Cephei is archetypical of Cepheid variables. In other words, it is a cracker of a variable star. It has a period of 5 days 8 hours and 37.5 minutes. The brightness rises rapidly to a maximum in about a day and a half and then it slowly dims over a period of 4 days. It has a prime location in the Northern night sky and its entire range of brightness can be seen with the naked eye. Using binoculars or a telescope, you can make out two stars at either side of it. One has the same brightness as the maximum brightness of Delta Cephei and the other has the same brightness as its minimum – so it is a good variable star to observe.

Goodricke is said to have used either very basic equipment or just his naked eye to make his observations, comparing how bright the star appeared against other stars in the night sky. He drew graphs of how the apparent brightness changed with time and used these graphs to determine the period of the variable stars.

With Cepheids, it is the star itself that pulsates like a beating heart, expanding and contracting on a regular basis. When the radius is the smallest and the density of the star a maximum, they expand rapidly. When the expansion rate of the star is the highest, the brightness and surface temperature of the star are the greatest. The compression of the star takes place over a longer period and this is when the stars are at their dimmest.

One of the amazing things about Cepheids is the regularity of their pulsations. They are so regular that we can measure the time taken for each pulsation to a fraction of a minute. You can almost set your watches by them so to speak. They become even more amazing when one considers the changes in size that are taking place in a matter of days.

Whilst Cepheids are fairly rare, there are still about 700 Cepheids in our galaxy; the Pole star 'Polaris' is a Cepheid variable. Its brightness pulsates with a period of just less than four days. However, the variation in brightness is small (about point one of a magnitude) so we do not notice it with the naked eye. Our own Sun is expected to turn into a Cepheid variable – but it will take another five billion years for it to do so!

All Cepheids are super giant stars - stars that are in the process of dying. Stars produce their energy by nuclear fusion. This is where two light nuclei 'fuse' together or, to put it another way, 'wham' into each other and 'stick' to form one bigger nucleus, producing heat energy in the process. At the beginning and middle of their lives, stars produce energy by the fusing together of Hydrogen nuclei i.e. protons (At these temperatures the

Hydrogen atoms have been stripped of their electrons), to form Helium - giving some energy in the form of heat as an extra bonus. To produce one Helium atom requires six protons initially, taken three at a time. Consider our three protons, two of them wham into each other but this combination of two protons is unstable so one of the protons decays into a neutron (by ejecting a positron) to give deuterium. Meanwhile, the third proton whams into the deuterium nucleus to give us Helium - 3 (two protons and one neutron). Whilst all this is going on, our other three protons have done exactly the same and so our original six protons have now combined to give us two nuclei of Helium - 3, which then wham into each other. In the collision, two separate protons are thrown out leaving us with a nucleus of Helium (two protons and two neutrons). The two free protons can go away to find four others and start the process all over again. However if we now add up the masses of the Helium nuclei and the two protons we find that it is less than the mass of the six original protons! Where has this mass gone? Well, Einstein had something to say about that with his equation $E = mc^2$. Mass and energy are equivalent and so the 'loss in mass' ended up as kinetic energy of the resulting particles and this is where the heat energy in the Star is produced.

These stars are stable, that is, they have a fairly constant radius due to the cancellation of the outward effects of pressure and radiation within the stars by the inward effects of gravity. Gravitational forces between the particles that make up a star try to make the star collapse. Pressure built up in the core of the star due to the intense heat along with the Star's radiation, tries to make the star expand. In 'stable' stars the two effects cancel and the star has a fairly constant radius and luminosity.

As a star gets older, all the Hydrogen has been used up in making Helium and they now start to fuse together the bigger Helium atoms. Once all the Helium has gone, the star fuses heavier and heavier atoms until it has used up all of its fuel. When there is nothing left to fuse together the star dies, often spectacularly. Cepheids have probably burned up all their Hydrogen and are now burning Helium in their cores.

The 'pulsating process' is something like this: When the star is large the inward gravitational forces are larger than the outward forces due to radiation and pressure and so the star contracts. In contracting, the gravitational potential energy of the particles making up the star reduces and is changed to kinetic energy making the atoms in the star move faster. Since the star is now denser, the chances of atoms bumping into each other and fusing together increases, which increases the rate at which heat energy is produced by nuclear fusion. Too much energy is now produced, the star heats up, and the outward pressure and radiation in the core is now stronger than gravitational forces trying to collapse the star. The outward forces are now bigger than the inward forces so the star expands and the accumulated energy is radiated outwards (the star is brighter).

In expanding, the particles gain gravitational potential energy and lose kinetic energy causing them to move more slowly. The density of the star also reduces and so the rate at which nuclear fusion takes place gets less. The core of the star cools down. The rate of nuclear reactions reduces, the star is dimmer, and the effect of gravity pulling the star inward is now greater than the effects of radiation and pressure pushing outward. The star contracts and we are now back to where we started. Hence the pulsating process continues.

These pulsations can only explain changes in brightness of up to a factor of fifty or so. Variable stars like Mira, the first variable star to be recorded, are much more dramatic and rather than just going bright then dim, more resemble Christmas tree lights which flash on and off. To flash on and off, these stars use the same process as that used in 'Sunblock' lotions to prevent us getting burned whilst out in the mid day Sun. As the star expands and cools down, molecules like Titanium Oxide and dust form in the outer layers of the star. Titanium Oxide is white and used in many sunscreens and paints. The oxides and dust absorb, or reflect back in, light trying to escape from the inner core of the star. Only light from the outer regions escapes but this radiation is in the invisible, Infra Red, region of the spectrum so we do not see it. Hence the star effectively disappears. As the star contracts and heats up, the energy of the particles is now too great for the molecular bonds to hold and so the Titanium Oxide disappears allowing the light to escape. This process causes the brightness of these stars to change by up to a factor of one thousand.

Likening Mira to Christmas tree lights is a good example as it has been proposed that Mira was the 'Star of Bethlehem', the star that appeared at the birth of Jesus and followed by the three wise men.

This is not as daft as it seems. The nearest star to us after our own Sun, Alpha Proxima, is just over 4 light-years from us. That is, it takes the light four years to reach us. Other stars in the constellation Ursa Major or 'Big Dipper' are about 70 light years away from us and so light from these stars takes about seventy years to reach us. Now if God was going to place a star in the sky to mark the birth of his one and only son, Jesus, and enable the three wise men to follow it to locate his birthplace, then he would have had to have pre planned the event beforehand by between four and seventy years or even longer. The reason for this being, that that is how long it will take the light to reach us from the new star.

Let's look at the logistics of this. God wants to place a new star in the sky to lead the three wise men from the East so that they can bring gifts to his newborn son and heir, Jesus. He wants the star to be an 'event' that will be recorded in history so it must be fairly close as the further away the star is, the dimmer it will appear on Earth. To appear in a miraculous 'flash' rather than a damp 'fizzle' and go unnoticed by the general populace requires this new star to be bright and therefore close to Earth. More importantly, he wants the star to be bright enough for the three wise men to be able to follow across the desert.

On the other hand, we don't want the star too close as it would be too hot down here on Earth and may singe all the inhabitants when it goes 'bang' - which might not be what God wanted. 'On the other hand' is a phrase used considerably when giving advice on what to do. Did you know that Arabic Sheiks always used to cut off the left hand of all their advisors? This is so that when they are giving advice the advisors cannot say 'but, on the other hand!'

Lets say, he will put the new star in the sky ten light-year away and so he must do this exactly ten years before the birth of Jesus. This is so that the light from this new star would travel through space and, ten years later; finally arrive at the Earth, exactly coinciding with the birth of Jesus. He would have also had to make a note in his diary when he created the new star to keep the night 9 years 3 months in the future free - as he would want to be immaculate that night for the conception.

It could be that way, but a variable star, switching on and off every 11 months could be said to be a little more probable.

But was Jesus born on the 25th December? Probably not as this date has an astronomical connection. It used to be the shortest day. People would celebrate and party all day long because the worst of winter was over. The days would become longer and warmer, crops would start to grow again and virtually everything anyone living in a mud hut could wish for was now on the way. This has now shifted to the 21st of December because the Earth wobbles on its axis (called 'precession'). It takes about 25 000 years to make one complete 'wobble'. However, since then, Gregory produced his 'Gregorian Calendar' that we use today and events such as this are now fixed and so the shortest day will remain on the 21st December from now on. So, whenever some spoilsport tells you to stop partying and to stop giving presents on the 'religious festival of Christmas', just tell them that you are not celebrating the birthday of Jesus who was born sometime else, but you are carrying on the tradition of celebrating the shortest day. Christians usurped this party so they could say that everyone was celebrating with them. But, let's get back to our story.

Goodricke's fascination with Delta Cephei proved to be his downfall. Whilst observing Delta Cephei he caught pneumonia and died at the age of 21years. His mum told him to wrap up warm but he didn't hear her. Even so, in his short life he was awarded the prestigious 'Godfrey Copley' medal by the Royal Society for his work on determining the period and cause of the variation in Algol and, two weeks before his death, he was elected a Fellow of the Royal Society. Not bad for someone whom society classified as mentally incapacitated because of his total deafness. It is not surprising that none of us students in Goodricke College, York University, knew who he was. In the College history, mention is only given to his explanation of the variation of Algol, no mention is made of the Cepheid variables which, in my humble opinion is far more important! There is a sculpture of Algol outside the College library.

Pigott went on to discover two more Cepheid variables and we will now go on to another story of discovery before returning to our route on the path of the Hubble constant.

Wait a minute! Did we say earlier that Pigott's constellation Antinous did not exist anymore and was now 'extinct'? How can a constellation become 'extinct'? Did it follow the Rolling Stones record and 'Just fade away'? Or did it go out with a bang? The answer is neither. 'Extinct' is the word used in astronomy books. To find out why it became 'extinct', you have to look elsewhere. Wondering just how a constellation could become 'extinct' I tapped in 'Antinous' into a search engine on the Internet. The results produced several promising sites and so I 'clicked' on the first one. A big sign appeared on the screen, 'Site blocked' - meaning that you are denied access to this site. This usually means that it contains something juicy. Now, I live in the Middle East and the Internet here is 'censored' by a 'proxy server'. Sites that are not considered suitable for people to view (i.e.'nooky' sites) are blocked so that the Internet can't lead one astray. Normally it works very well and it is a blessing in disguise as it prevents all of these unwanted 'popups' of porno pictures that seem to appear when using the computers in Internet cafes, when holidaying in Europe. But it has led to one or two problems. Anything with 'sex' in the header was 'blocked'. This immediately prevented anyone finding information on Universities in Essex, Middlesex, Sussex as these sites were 'blocked' by the proxy server because of their sexual connotation i.e. they all finish with 'sex'. The US government site on supernova was also blocked for a time as, for reasons best known to themselves, instead of the usual Internet address which starts http:/www... they had chosen http:/xxx.... and, being a triple x rated site, supernovae were also deemed as potentially ruinous to our moral well being and therefore 'blocked'.

So, lets rub our hands together in anticipation of why the constellation Antinous became 'extinct', and more interestingly, 'blocked'. The word 'Constellation' comes from the Latin meaning 'studied with stars'. A constellation is a group of stars that make a pattern in the night sky. The stars are not necessarily related (apart from all being within our galaxy - the Milky Way) and are not necessarily the same distance away. In fact the stars in a constellation are usually all whizzing off in different directions and the patterns we see now are not constant but changing. But, as most stars are so far away, it takes thousands and thousands of years before any noticeable changes can be seen. Additionally, the constellation Orion only looks like Orion from the Earth. If these same stars that make up Orion could be viewed from another part of our galaxy, it would look completely different. This is because the stars are so far away that they all appear to be in two dimensions – as if they were actually stuck on a celestial sphere. In reality the patterns are three-dimensional and so the stars would line up differently if viewed from somewhere else.

Who decided which stars were in which constellations goes back a long way. The Babylonians created some of the constellations (circa 5600BC) and some were even created before this. The Greeks didn't invent the constellations, but took the constellation shapes that already existed and named them after their own gods - making up their own stories, like the story of Andromeda and Medusa, to go with them (the Chinese did the same but their system never caught on in the Western world).

Now, Antinous is often known as 'the last of the Roman gods'. To find Antinous we must go back to the Roman Emperor Hadrian. Whilst on his travels, Hadrian met this young Greek lad (yet to reach his teens) and gave him a job in his household, educated him and eventually these two became lovers. It is not clear what Hadrian's wife thought about all this but I think we can guess (some think that she hated it as she might lose out on the money if Hadrian were to die and some think she was OK about it on the basis of 'one less job to do around the house'). In any case, Hadrian being Emperor didn't worry about this and took the lad with him wherever he went. On his travels to the outer reaches of the Roman Empire, Egypt, the young Antinous, who was now nearing twenty, foolishly fell into the Nile and drowned (was he pushed? We don't know, but one thing is for sure, wifey was back in the will, big time!).

Fortunately for Antinous, if you drowned in the Nile then the Egyptians believed that you became a god, so Hadrian, stricken with grief at the loss of his gay lover, proclaimed Antinous a 'god'. Temples were built, hundreds of statues of him were erected and one perk of being a 'god' was that you got to have your own constellation. That is what Hadrian did and why we had the constellation 'Antinous'.

So, Pigott was unlucky enough to discover a Cepheid variable, the first 'true one' of its class that was not only named after Antinous (Eta Antinoi) but it was also in a constellation named after Antinous. Both of these names had been given to honour the gay lover of the Roman Emperor, Hadrian. That is not a problem until you realize that that sort of thing was politically incorrect in the 17th and 18th centuries. You had to be careful where you pointed your telescope in those days.

Astronomers were now looking at the constellation Antinous, not with a view to studying it, but with a view to getting rid of it completely.

Astronomers (or more correctly, 'uranographers' - people who chart stars) were always coming up with their own ideas of which stars were in which constellation and sometimes naming them in order to gain favour with their patron. For instance, in 1624, Halley named the constellation 'Robur Caroli' in honour of King Charles II of England.

In 1690, a Polish astronomer had published a book containing a precise atlas of the sky. In this atlas, he tried to gain favour with his patron, the King, by naming a constellation in his honour ("Shield of Sobiescianum" or "Scutum Sobiescianum") and to do this he had to pinch a star from the constellation Antinous.

You can't go on drawing up and then re drawing constellations and having different star charts with different names or constellations on them. Astronomers use constellations as 'direction finders'. By looking for certain constellations you can find your way around the night sky, but not when they have no fixed boundaries or have different names depending on whose star chart you look at. Something had to be done. In the 1920's the International Astronomical Union (IAU) set the Frenchman Eugene Delaporte the task of standardizing it all. This he did and came up with the 88 constellations we know today. Antinous had to go. The constellation named after the Polish King, which included the star stolen from the constellation Antinous and designed to enhance the pension of the astronomer who named it, stayed – but with a name shortened to Scutum ('Shield'). To be fair, Eugene, a Frenchman, also dropped the constellation 'Globus Aerostaticus' the hot air balloon which had been named to honour the French Montogolfier brothers.

Astrologers, when 'working' out your horoscope, do not use this IAP standardized set of constellations – they stick to their old ones. Astrologers only have twelve constellations in their Zodiac whilst astronomers have several more.

That is why Pigott's constellation became extinct. To say that the constellation Antinous is 'extinct' is certainly the shorter way of explaining it – but less fun!

Chapter 2. Nobody believes a scientist!

For the next part of our journey we go back to 1519 when Ferdinand Magellan set sail to circumnavigate the World (I know it is before Goodricke and "when it all started" but be patient and you will soon see why). Magellan was born in Portugal at a time when great voyages of discovery were being made. If you were young and wanted to make a fortune in those times then you didn't deal in 'dot com ventures' as you do now, but you sailed off discovering new lands or trade routes. In those days it was spices that were the spice of life and where fortunes were to be made.

Portugal and Spain were the major international players at this time. Portugal had colonies in Africa and had travelled to India in the East. Portugal was becoming rich by trading in spices. Columbus had recently crossed the Atlantic and bumped into the 'West Indies' (thinking that he had arrived in India) and had claimed them for Spain, a country that was founding an empire in the West. But everyone knew the world was round. So somewhere, if Portugal kept going further and further to East and Spain kept going further and further to the West, these empires were bound to meet – and clash!

Things were hotting up between Spain and Portugal when the Pope (Alexander VI) stepped in to try and keep the peace between these two Catholic nations. Basically, in a series of letters or 'bulls', he divided the World into two halves, (from pole to pole) giving half to Spain and half to Portugal. Anything to the west of this line belonged to Portugal and anything to the east of this line belonged to Spain.

But of course, the Pope being a Spaniard himself decreed the Spanish 'half' to be bigger than the Portuguese 'half.' After a bit of 'discussion' the dividing line was moved so that it passed through the eastern tip of Brazil.

Naturally, it goes without saying that nobody new that Brazil or even the whole Americas existed at that time. Nor did anybody care what other countries such as Holland or Britain thought about it since they may have wanted a slice of the cake themselves.

The people who lived in these lands (known or unknown) were not taken into consideration either. The only exception was that if any land had a Christian king or power already ensconced before Christmas Day, 1493 then they had to be left alone. It must have been great to have the power to carve up the World.

Anyway, in 1511, the Portuguese had captured the City of Malacca. This was a city in Malaysia that everyone wanted as it was a major port on the spice route from the east to Venice. Anyone controlling Malacca could put a stranglehold on Venice's spice trade. And they did.

Now Magellan had disgraced himself in the Portuguese army. He had gotten himself seriously wounded whilst fighting in the Portuguese expedition to Morocco and worse still, he had a big falling out with his commander. So much of a falling out that he deserted the Army.

Magellan was a bit bored by now and was looking for some action. The Portuguese had 'discovered' the 'Moluccas' in 1512 (well, they had been there all the time actually). These are a group of volcanic islands in Indonesia. They are mountainous, fertile and humid and the original home of nutmeg and cloves. Other spices were also to be found there and for spices read 'money!'

A friend of Magellan wrote to him and told him about the discovery of the Moluccas and gave a brief description of where they were. Unfortunately, he over exaggerated how far east of Malacca the Moluccas were. This guy put them so far east that Magellan reckoned that they must be around the World, over the Pope's demarcation line and in The Spanish half of the World.

Off Magellan went to Spain and, to cut an ever lengthening story short, he took Spanish nationality, convinced the Spanish king to fund a trip to sail west and claim the Moluccas and its spices for Spain and return via the east becoming the first man to circumnavigate the world in the process. This, we are told, would also prove 'that the Earth was round'.

There are stories told by school teachers, and included in many text books (even in some science books I am sorry to say), that say that the church believed that the Earth was 'flat' and that anyone sailing to the edge would just 'fall off'. They go on to say that Magellan sailed around the World to prove that the Earth was indeed round and thereby prove the church wrong. We already know these stories to be incorrect since we know of the Popes 'Bull' dividing the round Earth into two halves but we will take a break in our story here to consider some science. You see the problem is, just nobody believes a scientist! In the 1500's, everyone knew that the earth was round and everyone should have known how big it was. The problem with navigators in those days was that they all underestimated the size of the Earth. Granted, finding longitude was a great problem in those days (an excellent story but one which does not concern us in our quest for the Hubble constant) but even so, navigators consistently underestimated the size of the Earth. This is why Columbus was convinced to his deathbed that having bumped into the 'West Indies' he had reached the West coast of India when he was still only half way around the Earth. It would prove the downfall of Magellan too, as we shall see later.

People have known that the Earth was round for thousands of years. They knew it was round because as they walked Northwards, stars in the northern sky appeared higher in the sky whilst stars in the southern sky appeared lower in the sky. Another reason that they knew that the Earth was round was in Lunar eclipses. Here the Earth's shadow falls on the Moon and the Moon enters a period of darkness. Now, the shadow of the Earth on the Moon is always round, never straight. The only thing that always casts a round shadow, no matter from which angle you look at it, is a sphere. They knew the Earth was round. But, one might ask, how often does the general populace watch the stars on their way home from the pub? Particularly as one would have to live a long way from the pub in order to see any noticeable difference in the apparent height of the stars in the sky as one walked home. Eclipses of the Moon are not that frequent either and blink and one might miss it. Strolling home discussing the problems of the Universe, exhaling noxious liquor fumes is not conducive to astronomical observations. No, astronomical observations like these are not going to get the rotundity of the Earth into the heads of the general populace. Ships coming into port though, would do the trick.

Up to the introduction of ship's radio around 1900, ships left port and they may or may not come back. If they did you were lucky, if they didn't you had no idea what had happened to them. They were 'lost'. Ships didn't sink because you just didn't know what had happened to them – so they were 'lost at sea'. As so many ships were 'lost' any boat coming home would make the owner or company owning it very rich indeed. Hence the phrase, I will buy you that expensive gift 'when my ship comes in'. If your husband or boyfriend was on board then you would be happy to see them return (if only for the money).

The general public watched out for ships.

Ships didn't just emerge out of a hazy distance but appeared bit by bit. First you saw the flag on the top of the mast appear over the horizon, then the sail on the mast and lastly, the hull itself would appear over the horizon. The ship arrived as if it were coming over a hill. The Earth was round. They knew it.

Why do you think Galileo (a few years later) was up at the top of the campanile (bell tower) in Venice showing all the big wigs his newly perfected telescope? When a ship came in, shares in that company hit the roof. With Galileo's telescope you could not only see the ships 'close up' but you could tell which ship it was hours before it reached port. This gave you insider information. You could nip down to the stock exchange and buy shares in the company that owned the homecoming ship (whilst everyone else was wondering whether the ship had been 'lost') long before the company shares skyrocketed. Galileo of course, made and sold scientific instruments including telescopes! It was only after this that he turned his telescope to the heavens.

If you are still not convinced that everyone knew that the earth was round think back to your school days and remember Eratosthenes. In the third century BC, Eratosthenes lived in Alexandria, Egypt and had a penchant with sticking sticks in the desert and measuring the lengths of the shadow cast. He stuck a stick in the sand at Syene (near Aswan now) at noon on June 21st (mid summer's day) and noted that there was no shadow. He did the same at Alexandria and found that the stick cast a shadow here. From this he postulated that the Earth is round and its surface is curved. By measuring the length of the shadow and knowing how far it is between Syene and Alexandria he was able to calculate the radius of the Earth and found it to be 6300km - differing by only a few percent from its actual value.

Now, if people thought the Earth was flat, what was old Eratosthenes doing measuring its radius in third century BC? Where did all this misinformation about ancient peoples believing that the Earth was flat come into it? To be honest, I don't know, but some say it is a story made up around 1830 by a Frenchman and an American. They were in Paris at the same time (but whether they met or not we don't know) and the American, one 'Washington Irvine' liked to write historical fiction and did so, including this story about the church not believing the Earth was round. The Frenchman, one Antoine-Jean Letronne is also said to have published the same story at a similar time in an anti-religious publication meant to cast a bad light on the church.

Why did navigators underestimate the size of the Earth? Because no body believes a scientist! We have already said that navigators in the 1500's consistently thought the world was much smaller than it really is. But in the 3rd century BC, Eratosthenes of Alexandria had measured the radius of the Earth to be 6300km and thus everyone knew that the Earth was round. The ancient Greeks knew how to calculate the circumference of a circle ('pi' or 3.142 times the diameter) so everyone (well anyone with an education) had known for over a thousand years how far it was around the Earth. Well everyone apart from Columbus and Magellan.

You can just imagine it can't you? A Spanish sailor saying to his wife,

"I've run out of tobacco dear. I'm just popping over to America for some more. Put the kettle on I'll be back in a minute, there's a good girl."

Well it wasn't quite as bad as that!

So where is Magellan on his voyage of discovery? Having received funding from the king of Spain, Magellan sailed off with five ships and 270 men across the Atlantic. Unfortunately, the Spanish captains didn't take to being led by a foreigner and three of the captains started a plot to kill Magellan - and this before they had even reached the Canary Islands! With one captain, Cartegena now held prisoner, they set sail for Portuguese territory of Brazil but they couldn't land there as they were sailing under a Spanish flag. Sailing down the coast of South America, looking for a passage through to the Pacific Ocean it became colder and colder and so they took shelter off Argentina. Unfortunately for Magellan, someone had released the mutinous captain Cartegena and he had started to plot another mutiny. Magellan put this mutiny down, stranded Cartegena on Argentina (Patagonia in those days) and either killed or imprisoned the other mutineers.

Eventually they found a passage through to the Pacific. Magellan christened this passage 'the strait of All Saints' but it is now known as the straits of Magellan. It took nearly forty days to sail through these straits and, at night, they would watch the Indian campfires burn. They called this land the 'Land of Fires' or 'Tierra Del Fuego'. During the passage, one of the ships turned back and went home. This was bad enough in itself, but unfortunately for Magellan he had placed most of the fleet's provisions on board this one ship and so it was a tragedy.

They eventually reached the Pacific and, or as Magellan thought, not far to the Spice Islands and riches beyond their dreams.

"Listen here, men. We are about to sail across this Pacific duck pond and reach the Spice Islands. Only a short hop, two, three days at most. No problem, hard bits over."

You see nobody listens to a scientist.

Four months later, what was left of the crew who had survived on rats, sawdust, putrid water and bits of leather off the sails arrived at the other side and bumped into the Philippines, which they claimed for Spain. Magellan then committed a cardinal sin for a foreigner; he got himself involved in local politics. Whilst trying to subdue one of the tribes, he and others were killed in battle by poisoned arrows. This meant that they did not have enough men left to man three ships. One was burned and the other two sailed on to the Spice Islands. Here they loaded up with valuable spices, but how to get home and avoid the Portuguese?

They decided to split up. One ship, the Trinidad, would return the way they had come whilst the other, the Victoria, would continue and return via India and thus circumnavigate the World. This plan was devised in the hope of ensuring that at least one of the ships would make it home. Unfortunately, the Portuguese captured the Trinidad and killed the crew. The Victoria returned to Spain and her sailors became the first to circumnavigate the World. It took three years for the journey, less than twenty men out of the original 270 made it back. But, the precious cargo of spices on this one boat more than paid for the entire trip.

So, what has all this to do with the Hubble constant? During the trip, one of the sailors, Antonio Pigafetta, kept a diary and fortunately for us didn't eat it during the short trip across the Pacific. Pigafetta hero worshipped Magellan and later published his diary to document the entire trip – and no doubt make some money. During the four months that they took to make the 'two or three day' trip across the Pacific and, having plenty of time since their days weren't disrupted by irritating things like meal breaks, they were able to study the heavens.

The night sky in the Southern hemisphere is completely different to that seen in the Northern hemisphere and you cannot see the Southern stars from the North. For instance, there is no pole star in the Southern hemisphere; it is just by chance that there just happens not to be a bright star in the correct place to mark it. Some pontificate that the absence of a Pole star explains why there is no bird migration in the Southern hemisphere or why no one in the Southern hemisphere is thought to have invented the wheel (the reason being that they never got the idea of everything turning about a fixed point).

What fascinated them though, were patches of starlight that looked like two large 'fuzzy white things', high in the night sky. At night and in good weather, these patches of starlight would produce a silvery shine on the ocean. What really fascinated the sailors was that they stayed in the same position in relation to the stars. The stars in the night sky are always in the same pattern. Orion is Orion; the Big Dipper is the Big Dipper - every night. They don't change their relative positions to each other as they appear to rotate from East to West in the night sky. The Earth spinning in the opposite direction causes this apparent rotation of the stars. Whilst the Moon and planets also follow this nightly motion, they move across the star pattern, being in a different position relative to the constellations every night. For this reason the ancients thought them to be 'gods' because they could 'wander' across the star patterns. The very word 'planet' means 'wanderer'. But these clouds didn't wander, they stayed with the stars as if they were stuck to them.

They were also big, the biggest objects in the night sky. The largest of the clouds is about ten times the diameter of the Moon. Pigafetta wrote about these clouds in his diary and named them after Magellan. He called them the Large Magellanic Cloud (LMC) and the Small Magellanic Cloud (SMC).

The first uranographer to include them in a star chart was Heveius in 1690, (published after his death). The naming survived the standardization of the IAU in the 1920's by the Frenchman Eugene Delaporte and are still known after Magellan.

We now know that the Magellan clouds are separate galaxies to our own, the Milky Way. They are companion galaxies to ours, close by, in astronomical terms, being a mere 200 thousand light year away for the LMC and a mere 170 light year away for the SMC. They are classified as 'irregular' galaxies i.e. they are fuzzy edged and contain a lot of gas, unlike the Milky Way, which is 'spiral' in shape. Several more companion galaxies to the Milky Way have been found recently and they are all said to be part of 'the local group'.

The Magellanic Clouds are still a very long way away but they are close enough to study. So close, that we can make out individual stars using the simplest of telescopes. Even more important in our quest for the Hubble constant is that they contain lots of Cepheid variables. This means that we can pick out an individual Cepheid variable star and watch how its brightness changes even though it is in a different galaxy to our own. This is just what Henrietta Swan Leavitt and the Ladies of

Harvard did and moved us on to our next step towards the Hubble constant.

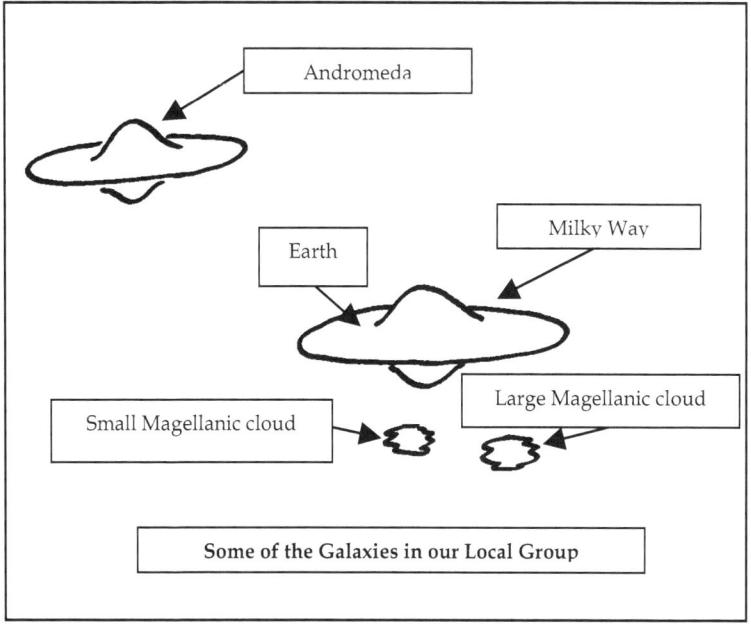

Some of the Galaxies in our Local Group

Chapter 3. Is there life on Mars?

The first problem in observing the Magellan Clouds, is that Harvard is in the Northern Hemisphere – at least it was the last time I looked, and the Magellan Clouds are in the Southern hemisphere and thus not visible to anyone at Harvard. To explain how this little problem was overcome, we need to go back a little and find two of the benefactors of the Harvard College Observatory.

Henry Draper was born in Virginia, USA in 1837. His father, John, was a doctor, chemist and professor at the New York University. Draper's father was interested in the 'new' photography and not only did he take one of the earliest portraits of a person in 1840; he also recorded the first daguerreotype (an early photographic process) of the Moon. As one would expect, his son, Henry helped in these photographic pursuits and became a keen amateur photographer and astronomer himself.

Henry followed the family tradition and went to medical school where he completed his studies early. Too early in fact, as he had to wait one year until he was twenty-one before he could receive his degree. Henry decided to take a year off and travelled to Europe where he visited the Lord Rosse's observatory in Ireland. At that time, this observatory had the World's biggest reflecting telescope, which boasted a seventy-two inch diameter mirror. He was so impressed by this telescope that he returned to the USA determined to incorporate photography into astronomical observation.

The use of photography in astronomy at this time was only beneficial in that it gave astronomers a 'hard copy' of what they saw to show other scientists. Until then, astronomers had had to describe in words, or draw pictures, of what they had seen. They had no real 'proof' other than trust in their own 'word' and the hope that other astronomers could repeat the observations and achieve the same results. At this time, the naked eye was still far superior to photographs as a measuring instrument.

On his return from Europe, Draper made a name for himself in the medical profession, married a wealthy society lady, built an observatory in the grounds of his parents' home and pursued his hobby of astronomy with his wife doubling as a hostess at the society parties by day and as a laboratory technician in the observatory at night. He was the first person to photograph the spectrum of a star, which he took of Vega in 1872, and he was the first person to photograph a nebula, photographing the Great Nebula of Orion in 1880.

On returning from a hunting trip to the Rocky Mountains in 1882, Draper became ill and died. His wife gave all of Draper's equipment as well as a large sum of money to the observatory at Harvard, with the request that it be used to continue Draper's work on incorporating photography into astronomical observations. Money was also donated to fund an award, which was named the "Henry Draper Medal" and was to be given for 'outstanding contributions to astrophysics'.

The director of the Harvard College Observatory was one Edward C. Pickering and, with the money from Henry Draper's widow, he instigated the "Henry Draper Catalogue." The purpose of this catalogue was to act as a complete source of reference, listing the spectra of all the stars. The first edition was published in 1890 and listed 10,000 stars, which was extended to 225,000 stars by the early 1920's. However, since the Harvard College Observatory is in the Northern Hemisphere, if the Henry Draper Catalogue were to be complete then it would have to include stars in the Southern Hemisphere as well. To do this, Pickering needed an Observatory in the Southern Hemisphere and this required the legacy of Uriah Boyden.

Uriah Boyden was a mechanical engineer from Foxborough, Massachusetts but wasn't what you would call a bundle of laughs. He was a vegetarian; he dressed in plain Quaker style clothing, and woke up before anyone else in order to stride out before breakfast, rain or shine, on his route march around town. Not only did he not drink alcohol but he didn't drink tea or coffee either. However, this did not prevent him from becoming successful as he completed several engineering projects, building railways and dry stone docks before moving on to building mills. He patented his own type of watermill and the income from this and several other inventions made him a very rich man indeed.

He died at the age of 74 in 1879 and left 238, 000 dollars with the express wish that it be used to set up astronomical observatories on the top of mountains, where atmospheric conditions would not influence viewing conditions. Pollution in the atmosphere hampers the observation of stars. The light that is not absorbed by the atmosphere is scattered making the image blurry. If you go to the top of a mountain, then not only are you away from sources of pollution (smoke, dust etc.) but also there is less atmosphere to look through so you get better viewing conditions.

This may have seemed a very reasonable bequeath if you were an astronomer, but if you were a family member, it didn't seem fair at all. Particularly as the money bequeathed represented just about all the money he had to leave! Eventually, the law courts decreed that he was of sound mind and so the will stood, the money was up for grabs and Pickering grabbed it!

Photography had come on a long way since Draper's early daguerreotypes of the Moon. Photographic emulsions were now much more sensitive which meant that exposures were much shorter. This led to additional advantages in the recording of astronomical observations photographically. One of these was the ability to take time exposures. It does not matter how long you stare up a telescope, you will still see the same stars. All that happens is that the longer you stare, the more tired your eyes become (which may well cause one to imagine things!) With photography and time exposures, the longer the exposure, the more light is collected from distant stars and this enables you to see very dim stars. It also enables you to 'see' stars that are otherwise too dim to be visible to the naked eye.

A second benefit is that with a photograph, you have a permanent record of your results, which can be studied later. This does not have to be in the observatory either. Photographic records of astronomical observations can be made in a cold and draughty observatory during the night and examined in a warm laboratory at leisure. Electronic imaging and the Internet have made life cosier still for the couch potato astronomers of today.

They can control a telescope remotely over the Internet and have the images delivered to their computer for analysis. The telescope can be, and often is, at the other side of the world, or even in space where viewing conditions are at their best.

Pickering decided to use the money from the Boyden estate to set up an observatory on the top of a mountain in the Southern hemisphere. Peru was chosen not only because the climate appeared advantageous for observing but also because Peru is due south of Harvard. This meant that it would not be too far away from Harvard as travel was still a problem, and the intention was for photographic observations to be made with the photographic plates being sent back to Harvard for analysis.

Solon Bailey was chosen as the man to lead the expedition to Peru to find a suitable site. Along with his wife and his young son, they went to California and observed the solar eclipse of January 1889 before setting sail from San Francisco. Bailey describes the History of the expedition in the Harvard College Observatory Annals in Chapter 1 of his scientific paper (34:1-48), before going on to give the results of their observations. It seems strange to us today to have a travelogue describing the weather, the scenery and the people they met published in a scientific journal. The Baileys sailed towards the final port of Panama on a mail boat, which called in at various ports on the way. He describes Acapulco in Mexico as "not now a place of such importance as formerly" but he was most impressed by the horseshoe harbour.

I too was impressed by the harbour when I spent several weeks there myself - but in my case I viewed the harbour from the bars overlooking the bay that lined the cliff tops. I mention this as it was during my time in Acapulco in 1999, both in my hotel room and whilst sampling the fine Mexican ales in a bar named after the pirate Barbarossa (who was based in Acapulco) that I finally succeeded in performing the calculations for the Hubble constant and the exponential form of the redshift distance relationship that we will meet later.

The Baileys sailed on to Panama where they were met by Solon's brother Marshall who arrived direct from New York with one hundred cases of equipment. From Panama they travelled by train, crossing the as yet unfinished Panama Canal. In fact the de Lesseps company had stopped work and Bailey, stunned by the size of company's vast hospital and burial grounds, was convinced that work on the canal would "definitely be abandoned".

Having successfully built the Suez Canal and made vast sums of money, de Lesseps had started on the Panama Canal with a view to making even more. Unfortunately, the region around Panama was plagued with mosquitoes but at that time, no one knew that mosquitoes transmitted both malaria and yellow fever; they thought that both diseases were caught from 'invisible mists' emanating from decaying matter. The name 'malaria' comes from the Italian for 'harmful air'. Work on the canal carried on regardless and over thirty thousand workers died of one or other of the two diseases. Bailey was partly correct in that de Lesseps stopped work and the company went broke, but the building of the canal resumed under the direction of the US Government. They were aided by the fact that a Cuban physician had identified mosquitoes as the carriers of the two diseases and the realisation that if the mosquito breeding grounds were removed, so too was the problem.

The Baileys carried on to their final destination of Chosica, a town not far from Lima, taking nearly two months for their journey. Chosica is a town near Lima, the capital of Peru, and it had originally been chosen because it was in the Andes and what little weather information was available indicated that it could be suitable for their observations. This was not the case though, as on arrival they discovered that to the South were very steep mountains, which blocked out their view of the Southern horizon and would prevent their observation of many of the Southern stars.

Aided by the superintendent of the railways, one Mr. R. B. Hubbell (was this an omen of things to come?) who provided them with a guide, they went out on scouting missions to try to find a new site. They trekked by mule and foot over the mountainous and rugged terrain, finally finding a mountain that they thought would be suitable for their needs and named it "Mount Harvard". Whilst it may have looked a suitable site for astronomical purposes, being completely barren with no water or local food supplies it was hardly a suitable place to live. They decided that they would set up the observatory there on a temporary basis and begin their observations, even though everything would need to be brought to them by mule train and then look elsewhere for something better once the cloudy season began. It must be remembered that water was not only needed for them to drink but large amounts would be needed to process and wash the thousands of photographic plates that they intended taking and this too would have to be brought by mule train. They had brought two small portable houses with them. These consisted of wooden frames, canvas roofs and "building paper on the roll to cover the walls."

It soon became clear to Solon that the site on Mount Harvard was not suitable and that they would have to move on. Too often high clouds interrupted their observations and so they needed a site with better weather conditions. Whilst they were in Lima they had realized that no one had ever taken methodical records of rain, cloud, or other atmospheric conditions and so they had to rely on 'word of mouth' in finding a suitable site. But the problem was that different people gave different advice. Arequipa was one place suggested as having excellent weather whilst an English meteorologist, Mr. Ralph Abercrombie, 'spoke very highly' of the Atacama Desert, which was much further South in Chile.

Leaving staff at the observatory to try and make whatever observations they could (which proved to be almost none because of torrential rain), Solon and Marshall set off on a four-month journey to look for a better site.

Both Arequipa and the Atacama Desert seemed appropriate places to relocate the observatory with the Atacama Desert being the most promising from an astronomical point of view. Solon thought that the place was far too desolate to be able to run an observatory there and so decided against the Atacama Desert. Interestingly, this is where most of the Earth based big telescopes of the present day are now located (apart from the Boyden telescope, which was eventually moved to South Africa), the Internet having overcome the practical difficulties, allowing the telescopes to be operated from anywhere in the World.

Pickering chose Arequipa and instructed that the observatory be moved there. Furthermore, Pickering decided that the Baileys had endured enough over the two years and so repatriated them, replacing Solon with Pickering's own brother William. William Pickering was a competent astronomer and is now credited as being the first person to discover a satellite by photographic means since he went on to take photographs of Saturn's moon, Phoebe in 1899. His brother Edward had given William instructions to the keep all costs down as they may find Arequipa unsuitable and may have to move on elsewhere. In total disregard of these instructions, William purchased a plot of land on which to build and spent so much money on making life comfortable for his family that he went nearly twenty times over his initial budget. Not only did William disregard his brother's instructions regarding expenditure, he also took no notice of his purpose for being at the southern observatory. Within the first year that the observatory was operational, hardly any photographs of the Southern skies had reached the team at Harvard. Orders, demands and pleas went unanswered as William was bent on following his own line of research.

In 1877 an Italian astronomer, Giovanni Sciaparelli, observed Mars at its 'opposition' when viewing conditions are at their best and reported seeing linear channels on its surface. Mars at 'opposition' means that the Sun, Earth and Mars are in a straight line, and in that order. Since the Sun and Mars are on opposite sides of the Earth, we have the name, opposition. Since the Earth is closer to the Sun than Mars, at opposition, not only is Mars fully lit but also, it is very near in astronomical terms to the Earth. The Italian word for 'channels' is 'canali' and people instantly thought that he meant 'canals'. Now canals are man made and so this meant that there must be 'life on Mars'. Most professional astronomers were sceptical of these claims but the more Sciaparelli observed Mars the more sophisticated and prominent the 'canali' became. The number of canals seemed to change on a seasonal basis and since several other reputable astronomers also reported seeing this canal system, the alien life on Mars story stayed in the news for some time and caused quite a controversy. It even prompted a spurt of science fiction books on alien life including "War of the Worlds' by H. G. Wells. It is hardly surprising then that William ignored the mundane task of taking thousands of repetitive photographs of the Southern skies just in order to catalogue them and turned his telescope onto Mars instead.

William reported seeing forty lakes on the surface with clouds high in the Martian sky. He proposed that the canals were made visible by strips of vegetation at either side of the canals and that is why the number of canals seen changed with the seasons. Percival Lowell used these observations and others to develop a theory of an advanced civilization on Mars. The canal system carried water from the ice caps of Mars to great cities near the Martian equator. William Pickering fully agreed with Lowell's hypothesis, in fact William also believed until his death, that there was life in one of the craters on the Moon.

What upset his brother Edward the most, was that he and the team at Harvard only came to know about his brother's exploits by reading reports about the Martian 'discoveries' in the popular newspapers. Instead of reporting his findings in the academic press, William was sending telegrams direct to the 'New York Herald' and rather than using the normal, cautious, scientific language he was reporting his sightings on Mars as proven fact.

Edward felt that this was bringing the Harvard College Observatory into disrepute (I can't think why!) and ordered his brother home. This was not an easy task as William refused to return, as he wanted to continue looking for life on Mars. Eventually, Solan Bailey, the original founder of the Southern observatory, was set the task of cataloguing the Southern stars and restoring the reputation of the observatory by replacing William Pickering. In the preface to the scientific report furnishing magnitudes for the Southern stars, Edward Pickering pointedly remarked, "all of the observations were made by Professor Solon I Bailey and nearly all were recorded by Mr. Marshall H Bailey". No mention is given to Pickering's brother, William.

Because of the sheer scale of the operation, Edward Pickering had to automate the whole process and to do this he employed a corps of women. He gave credit to them by stating "Important aid was rendered throughout this work by Mrs. M. Fleming, and the large amount of clerical and numerical work was done by the corps of female computers under her direction". Calling the women a type of machine, 'a computer' was not meant as an insult. At that time, anyone who was employed to perform calculations was called a 'computer' – the electronic machine was named after them and not the other way around. Other women were hired to record the information and these were called 'recorders'. These computers and recorders were paid thirty to fifty cents an hour, which was more than a factory worker but less than clerical office staff.

All the photographic plates from both the Northern and Southern star surveys were sent back to the Harvard College Observatory for analysis. Allegedly, Pickering originally had a male assistant to help him, but he was unhappy with his performance and in a moment of anger, told him that his maid could do a much better job and set her to the task just to prove it. Despite not having had any training in mathematics or science she proceeded to perform the task much more efficiently than the male assistant had done earlier. Pickering gave her the job permanently, there and then. Five years later, the maid, Mrs. Fleming was promoted and put in charge of the Henry Draper catalogue becoming the 'Curator of Astronomical Photographs', the first 'proper job' given to a woman at Harvard.

One of the female computers working under Mrs. Fleming was Annie Jump Cannon. Annie was born in Delaware in 1863, daughter of a wealthy shipbuilder and state senator. She was hard of hearing but despite this she studied physics and astronomy. After graduating, she returned home to be with her family. As well as being 'the dutiful daughter' Annie loved to travel and recorded her adventures photographically. After the death of her mother, she returned to her studies becoming a 'special student' of astronomy at Radcliffe. In 1896 she was hired by Edward Pickering to work on the Henry Draper catalogue classifying southern stars according to their spectra.

Whilst we might think of stars as being 'twinkly' white, they actually have different colours. The 'colour' white is not actually a colour but a mixture of all the colours in the rainbow. A 'red' star mainly emits colours from the 'red' part of the rainbow or, to give it its proper name, spectrum. A blue star will emit colours mainly from the blue end of the spectrum and little from the red end. As to what colour a star is depends upon the age of the star and its temperature. The youngest and hottest stars appear as blue-white whilst the oldest and coolest appear red in colour.

Annie Jump Cannon devised a way of classifying stars according to their colour and hence temperature and age. She divided the stars into seven types O, B, A, F, G, K and M. The letter 'O' represented the hottest, blue-white stars whilst the letter 'M' represented the coolest, red stars.

Students of astronomy remember the sequence by the mnemonic, Oh! Be A Fine Girl. Kiss Me! Just as when white light passes through a glass prism the light is dispersed and splits up into its component colours, the same happens with the light from distant stars. A glass prism was attached to the telescope (or better still, a diffraction grating) and the spectra recorded on photographic plates. Annie would then classify the stars according to their spectra.

Between 1911 and 1915 she classified nearly a quarter of a million stars by their spectra, that is 5000 stars per month! Whilst someone stood behind recording the information, Annie would study the photographic plate, determine the spectral class of the star and call out the alphabetical name of the class, which would be recorded by the recorder. Working in this way, she could classify as many as three stars a minute.

What was more impressive was the reliability with which she could classify the spectra. In science, reliability of experiments and procedures is displayed in our ability to repeat them over and over again and still get the same result. Annie could repeat the classification of a star several years later and still give an answer to within one tenth of a category's subdivision.

Since there were so many stars to classify, speed was essential and every minute counted. If there was a delay of just one minute in the classification of each star, then the publication of the whole catalogue would be delayed by as much as two years. The original Henry Draper Catalogue was published between the years 1918 and 1924 by Anne Jump Cannon, (who was then curator of astronomical photographs since Mrs. Fleming had retired at that point) and Edward Pickering, Director of Harvard Observatory. Astronomers finally had a comprehensive source of reference on the magnitudes and spectra of stars in both the Northern and Southern hemispheres of the sky.

Chapter 4. Blackpool illuminations.

If you are still lying on the sofa pondering the Universe (as you may have been doing, when beginning the book), you will remember that as we climbed the ladder in order to change the bulb in the ceiling light, the bulb became bigger and bigger, as we got nearer and nearer to it. Well, there is another change associated with the distance from a lamp and that is; the closer we are to a light source, the brighter it appears to be. Having replaced the lamp and asked someone to switch on the electricity in order to check that the lamp is now working, we see the lamp becoming dimmer and dimmer as we retrace our steps down the ladder and back to the sofa.

This is because light sources emit light equally in all directions. Imagine that you have become bored with the old tiles on the bathroom wall and wish to brighten them up by painting them a different colour. An easy way to do this is to use an aerosol spray to paint the tiles. As the spray paint leaves the nozzle and travels further and further, the globules of paint spread out so that the area covered by the paint becomes larger and larger.

Let's take an example. You find that if you hold the can of paint 10cm from the wall, the stream of paint has spread out just enough to cover one tile on the wall. One minute later you have a perfectly painted tile with a decent thickness of paint on it. But there are an awful lot of tiles on the wall and it will take forever to paint all the tiles at this rate. It is at this point that you are struck by a brilliant idea.

"Why don't I move the can of paint back twice as far so that it is now 20cm from the wall?"

The stream of paint will spread out so much that it will be twice as deep and twice as wide. The area of coverage will now be four times what it was before. Instead of just painting one tile at a time, you can now paint four tiles at once!

However, when you try this, you find that whilst the paint does cover four tiles at the same time, it now takes you four times as long to paint them. Instead of taking one minute to paint one tile it takes four minutes to paint all four tiles - so you are no better off! The reason for this is that it is the same paint that arrived at the single tile 10cm away from the nozzle, that is now arriving at the four tiles 20cm away. The amount of paint arriving each minute on every tile is now one quarter of what it was before. The stream of paint is less intense because it is more spread out. If you go twice as far away, the paint is spread over four times the area and the intensity falls to one quarter of its original value. This is known as the inverse square law.

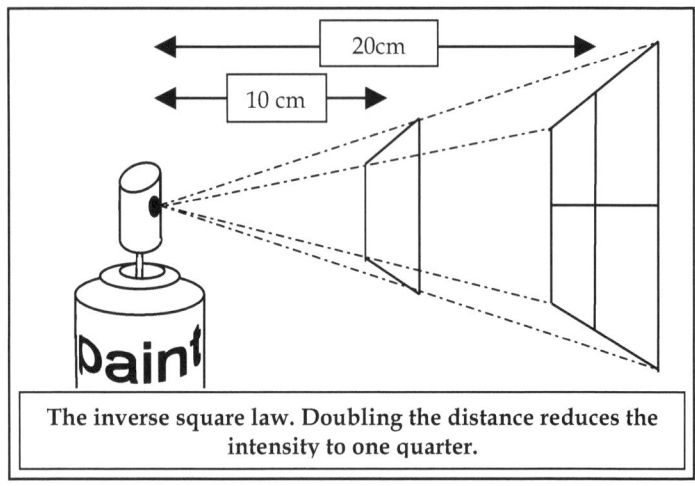

The inverse square law. Doubling the distance reduces the intensity to one quarter.

It is the same with light travelling from a distant star. The light emitted by the star spreads out equally in all directions. If a star is twice as far away then the light emitted by it will have spread out so much that it now covers four times the area thus appearing to have only one quarter of the original brightness. The problem is, when we look at stars in the night sky, some stars appear to be bright and some stars appear to be dim. This difference could either be because these stars are intrinsically brighter (that is actually brighter) than other stars or it may be because the stars are the same brightness as each other but that one star is further away from us than the other. You just don't know.

Astronomers use specialist terms to compare brightness. A reader who is an astrophysicist will already know these terms. A reader who is not an astrophysicist doesn't need to know them.

Consequently we will not go into these terms in any detail but we will restrict ourselves to just two of them. These are the 'true brightness' and the 'apparent brightness'.

How bright the star really is, we will call the 'true brightness' (also known as the luminosity) of the star. How bright the star appears to be when we view it from the Earth we will call the 'apparent brightness' of the star.

The true brightness of a star is independent of the observer; it depends only on features of the star such as surface temperature, radius, composition and so on. The apparent brightness depends upon the observer, such as how far away we are from the star or how much dust there is between the star and us.

If one thinks of a one hundred watt light bulb, its true brightness will always be equivalent to one hundred watt but its apparent brightness, how bright it appears to be, will depend on how far it is away from you.

When I was a lad, my father used to take me to Southport to look at Blackpool illuminations. This may seem a little strange, as normally people go to Blackpool to see Blackpool illuminations, but we will not go into that. Blackpool and Southport are both seaside towns on the West coast of Lancashire, England separated by the estuary of the river Ribble, which flows into the Irish Sea at this point. Blackpool is the original Las Vegas, a town built solely for the purpose of having fun and it was the first town in Britain to install electric street lighting. In 1879, eight electric arc lamps were installed with the result that over one hundred thousand tourists came to Blackpool to cavort underneath the glow from the lamps. This was no mean feat, as remember there were no cars and motorways in those days; only trains and horses. Realising that they were on to a good thing, every year Blackpool now puts on a magnificent display of 'illuminations' to attract tourists. The whole six miles of the promenade is 'lit up' by electric lights of every description. Southport, on the other hand did not.

Standing on the sea wall at Southport, one could look across the Ribble estuary and see the illuminations strewn along the length of Blackpool promenade. Since all the light bulbs were the same distance away from us when we stood across the river in Southport, lights that appeared brighter actually were brighter. Looking from a distance, it is easy to compare the true brightness of one light bulb with another because the distances involved are the same. If the lamp appeared twice as bright to us in Southport, then it truly was twice as bright in Blackpool. Or to put it another way, a two hundred watt lamp looked twice as bright as a one hundred watt lamp, whether it was viewed from Blackpool or from Southport.

If we looked out to sea then it was a different matter. Here we could see some ships and some had bright lights on board; some had dimmer lights on board but you could not compare the true brightness of the lamps. The ship with the lights that appeared brighter may either have been closer to us than the other boat or it might simply have had brighter lights on board. There was no way of telling. If we knew that the ship had 'one hundred watt' lamps on board and that the lamps in Blackpool illuminations were one hundred watt, then we could compare the distances by comparing how bright the lamps appeared to us in Southport. If the apparent brightness of a 100W lamp in Blackpool has four times the apparent brightness of a 100W lamp on the ship, then we would know that the ship is twice as far away from us as Blackpool – remember the inverse square law. If we know how far Blackpool is away from us in miles, then we can easily find out how far away the ship is. We call the 'one hundred watt lamp' here a 'standard candle' because it is a 'standard' by which we can compare distances. A 'standard candle' is a light source that has the same true brightness wherever it is in the Universe.

To solve our problem of how far away stars are, what we need is a "Blackpool illuminations" in space! That is we need a set of stars that are to all intents and purposes, the same distance away so that we can compare their brightness. This is where the Magellan Clouds enter the picture; they are our Blackpool illuminations in space! The Small Magellanic Cloud is approximately five thousand light years across and one hundred and seventy thousand lights year away. Whilst some of the stars might be one hundred and seventy thousand light years away and some one hundred and seventy five thousand light years away, to all intents and purposes, the stars are all the same distance away. It is like saying that the distance from London to New York is three thousand miles. We do not worry if we mean the distance from Mayfair to Wall Street or from Oxford Street to Central Park; to all intents and purposes the distance is three thousand miles. It is the same with the Small Megallanic Cloud; all the stars in them are the same distance away from the Earth.

Henrietta Swan Leavitt worked at the Harvard College Observatory and was set the task of studying the variable stars in the Small Megallanic Cloud. Henrietta was born in Cambridge, Massachusetts, in 1868. Born to a Congregational Minister, she went on to study astronomy at what is now, Radcliffe College. After graduating she became seriously ill and lived at home for many years. On recovering from this illness, which had left her deaf, she took a voluntary post on the staff at the Harvard College Observatory, working under Edward Pickering. It took a further seven years before she was appointed to the permanent staff at the usual salary of thirty cents an hour.

She was not allowed to follow her own research project; she had to do what Pickering told her. He set her the boring task of cataloguing the variable stars in the Magellanic Clouds but we have to remember that women were not allowed to work at night or go near the machinery of a large telescope. This posed quite a problem for female astronomers and cataloguing stars was probably as close as she could get to 'real' research in astronomy.

Identifying variable stars in the Magellanic Clouds was very difficult, as they had to be identified from photographic plates. Since the stars were so far away they were very dim and this required long exposures. This meant that photographs taken at different times to show if the brightness of the stars was changing, showed little difference between them as the brightness of the stars changed during the exposures and what was produced on the photographic plate was an average of the apparent brightness of the stars during that exposure. Long exposures also meant that not many exposures could be taken in a night so she had very few photographs to compare.

Leavitt formulated her own techniques for identifying variable stars. She took two photographs of the Small Magellanic Clouds taken at different times. One she left as a negative showing the stars as black against a clear background. The other she made into a positive, giving clear stars on a black background. When the negative image was superimposed over the positive image, non variable stars disappeared, as the black on one plate cancelled with the clear on the other. Variable stars did not cancel as the brightness had changed between exposures. Difficult as it was to identify variable stars, she still managed to find 1,777 variable stars in the Magellanic Clouds. She then went on to measure the period of twenty five of these variable stars in the Small Magellanic Cloud along with their apparent brightness. Since they were very faint, it was impossible to measure their spectral type but she noted that they all resembled the variable stars found in globular clusters. That is, they dimmed slowly, remained dim for the greater part of the time, and then their brightness shot up to a brief maximum.

Leavitt noticed a pattern in the results. The brighter the Cepheid variable then the longer the period over which it varied. In other words, A Cepheid variable whose brightness varied over a period of five days had a greater apparent brightness than one whose brightness varied over a period of two days. Since they where the same distance away, this meant that this relationship also applied to the true brightness of the stars. By comparing the period of Cepheid variable stars, Leavitt had found a way to compare their true brightness's. However, the brightness of a variable star changes, so what do we mean by 'true brightness'? With variable stars, once we have determined the period, we take the average brightness for our calculations.

Here we have a way to measure distances in space. Find a Cepheid variable in a distant galaxy. Measure the period and the average apparent brightness of this variable star in the distant galaxy. Look at the Small Magellanic Cloud and find a Cepheid variable that has the same period as the one in the distant galaxy. Measure the apparent brightness of this star. Since both stars have the same period, they must have the same true brightness. Using the inverse square law, if the Cepheid in the distant galaxy has only one quarter of the apparent brightness of the Cepheid with the same period positioned in the Small Magellanic Cloud, then the galaxy must be twice as far away. If the Cepheid in the galaxy has only one ninth of the apparent brightness of the Cepheid in the Small Magellanic Cloud the galaxy must be three times as far away as the Small Magellanic Cloud. Leavitt derived a formula whereby the apparent brightness of any Cepheid variable in the Small Magellanic Cloud could be determined, once the period was known.

The trouble was that they did not know how far away the Magellanic Clouds where! It was like having a map with the corner missing and on this corner was the scale of the map. From your defective map, you know that Glasgow is twice as far from London as Birmingham is, but that is all. Without the scale, you have no way of determining the actual distances. If we could find just one Cepheid variable close enough to us, to find how far away it is by the method of parallax, then we could calibrate our distance scale and restore the corner of the map but, unfortunately, there are no Cepheid variables close enough to measure their distance by parallax. Also, the relationship only holds for a particular class of variable star, so as far as finding how far away things are in space, we are no nearer! It was still a major breakthrough though and had Leavitt not died of cancer at the age of 53, she may well have won a Nobel Prize for her discovery. As it was, she published the paper (or should we say E. C. Pickering published the paper 'for' her – let's not get silly, she was a woman after all!) in March 1912 in the Harvard College Observatory Circular, with the net result that the Titanic sank in the following November!

Ejnar Hertzsprung is credited as being the first person to measure the distance to the Small Magellanic Cloud - which he did in 1913.

Have you ever wondered why famous scientists have such strange names? I mean, why hasn't anyone just 'ordinary' ever invented or discovered anything? So far we have had Goodricke, Pigott, Swan-Leavitt to name but a few and now we have 'Ejnar Hertzsprung'. Let us look at this statistically. Most people in the World are supposed to have the surname 'Smith' (or some overseas version of this name) and so most of the discoveries should have been made by someone called 'Smith?' Smith's laws, Smith's gadgets or Smith's principles should inundate us, but they do not. Does this mean that if one is born with a common name then one is not as innovative as anyone else or are people with common names conditioned into thinking that they are 'not special'? I appreciate that when we are looking for something, we can always find the things that we do not want rather than the things that we do want. This is because there are more things that we do not want and so they are far more likely to be found. For instance, when one looks for a particular sock, we can usually find every sock we have ever owned before we eventually (if ever!) locate the errant sock that we are looking for. Since there are more socks that we do not want, it follows that it is more probable that we will find a sock that we do not want, and that is what happens. Even so, there are not that many scientific principles that have been discovered by people with common names. This has nothing to do with our story and I am not trying to score points; after all, Ashmore is a fairly common name in itself, but maybe someone should investigate the phenomenon of why so few discoveries have been made by the majority of people – who have common names.

As a student, Ejnar Hertzsprung never attended an astronomy lecture in his life, and yet he became a professor in astronomy and an observatory director. He was born in Copenhagen in 1873 and went on to study chemistry before taking up a post in St. Petersburg, Russia, researching into acetylene lighting. When his brother died, he returned to Copenhagen to look after his sister and mother. Being from a rich family, he lived the life of a 'gentleman scientist' and out of interest, started visiting the local observatories. He carried out his own private research into photography and only later turned to astronomy. Hertzsprung, realised that there was a relationship between the spectrum of a star and its true brightness. He published his findings in a photographic journal - with the end result that no one working in astronomy saw it. Fortunately, in 1909, he was working as an astronomy professor in Potsdam and one of his colleagues, Schwarzchildt, travelled to America where he bumped into H.D.

Russell (I suppose that Russell is a fairly common name). Russell, working independently in America and not having read many photographic journals, had come up with the same results as Hertzsprung and not realised it.

They looked at nearby stars; ones whose distances could be measured by parallax and determined the stars apparent brightness measured here on Earth. Knowing how bright they appeared to be and how far away the stars were, it was then a simple matter to calculate the stars true brightness. When they plotted the stars' 'true brightness' against the spectral classes of the stars, as classified by Annie Jump Cannon, they found that the points all lay in a pattern similar to a reversed number '7'. Any star lying on this reversed '7' is known as a 'main sequence star' - which is a bit of a misnomer as it implies that the stars gradually move along this path as they age, when in fact, they tend to spend most of their lives at the same stage on the diagram. This scatter graph is known as the Hertzsprung-Russell diagram (HR diagram for short) on the basis that Hertzsprung discovered it first. This diagram can be used to measure distances in space. Find a star, measure its spectral class, and the HR diagram will tell you its true brightness. Compare this with its apparent brightness here on Earth and one can calculate how far the star is away. The only trouble is that it is not very accurate when applied to a single star. On the HR diagram, stars lie within a band and not on a single line. Consequently, there is not a single true brightness for each spectral class but each class of star has a small range of true brightness that these stars can have. To put it another way, it is uncertain as to what the true brightness of a star is from its spectral class; only an approximate value can be found. These errors are compounded as distances to galaxies further and further out into space are measured using the distance ladder, where one method builds upon another. Any small errors in distances here are magnified as we go further and further out into space.

In 1852, the good townsfolk of Albany, New York decided that they wished to have in their midst a World class observatory and thus set about raising sufficient funds to build this scientific edifice by seeking donations from the community. The widow of a Senator Charles Dudley made by far the largest donation and was rewarded by having the observatory named the "Dudley Observatory". A governing body was set up to manage the observatory and its members were drawn from respected public figures within the community. Since these people were basically amateurs, a Council of highly respected scientists was enlisted to form a group who would oversee the research programme and make the Dudley Observatory a household name throughout the World. This Council had Benjamin Gould as its leader along with Benjamin Pearce, a mathematician who had helped set up the Harvard Observatory.

It is interesting to note that Harvard University had just made a major contribution to the US legal system by providing the first piece of forensic evidence to be accepted by a court. One of Pearce's colleagues, a Chemistry professor named John White Webster, borrowed some money and provided a collection of rocks as collateral. However, the naughty boy had already used the rocks to guarantee a previous loan elsewhere. An argument broke out with the result that the professor murdered the lender, cutting him into pieces and hiding the 'bits' around the Chemistry laboratory. It was only when a janitor smelled something 'fishy' and started to knock holes in the walls of the lab that the various body parts were found and thus the murder was discovered. The Professor was found guilty and hung in 1850. What about the forensic evidence? Well, the victim's false teeth were presented to the court to prove the identity of the victim and became the first piece of forensic evidence to be accepted by a US court. But, let us return to the story.

With Gould as Director, who better to make a name for the Dudley Observatory, since he was the first American to receive a Doctorate for astronomical research and also the founder of the Astrophysical Journal? Thus the future of the observatory was assured. With local dignitaries in charge of the finance and the management of the observatory and prominent scientists to direct the research, what could possibly go wrong?

Gould had been sent to Europe to purchase the telescopes and other astronomical apparatus and this had arrived back at the observatory for assembly. Although Gould gave most of his instructions by correspondence, all was going well until the project ran behind schedule and funds started to run a little low. The governing body was not deterred and wanted to move forward by hiring a full time member of staff as director but the scientists were against this, as they wanted to take their time and get things right.

A difference of opinion arose and, just as oak trees grow from little acorns, the minor tiff developed into open warfare. Tempers flared, discussions grew into verbal slinging matches, the governing body shouted insults at the scientists, and the scientists slammed the Observatory door in their faces and locked it. No amount of shouting through the keyhole by the governing body could tempt the scientists to open up and let them in. The governing body felt that they had a duty to the community who had raised the funds to manage the Observatory and see that the funds were properly accounted for. The scientists grew arrogant and decided that since this was to be such an illustrious Observatory of National fame, it should not be run by a bunch of amateurs like the board, but should be a 'monument to scientific research'. Verbal rows led to letters and articles in the local newspapers, which led to scandalous reports in the national press. Eventually, the siege was ended, Gould was evicted and the Governing Body won but in war, nobody really wins. Gould went on to be the first to use the newly laid transatlantic cable to measure longitude and spent nearly twenty years in Argentina cataloguing the Southern skies before returning to resume publication of the Astrophysical Journal. The Observatory on the other hand, was shunned and it would be twenty years or more before it was awarded any sort of respected research project.

In 1876, Lewis Boss was appointed as director of the observatory and that is when things started to pick up. Boss was not only an excellent astronomer but he was also a political animal who could tread the fine line between the Governing Body and the needs of the scientists. Lewis and later, his son Benjamin, directed the Observatory for over eighty years and during this time produced several star catalogues, giving precise star positions. More importantly to us, they also catalogued 'proper motions' of nearby stars. That is they determined the direction in which nearby stars were moving and also, the angle through which they moved each year. Lewis went one step further and realised that all the stars in the Hyades cluster were moving towards one 'vanishing point'.

The Hyades is a cluster of over one hundred stars, about one hundred and fifty light years away. They are all the same age and are held together by their mutual gravitational attraction. With the exception of the bright red giant, Aldebaran, they form the 'V' shaped head of the constellation Taurus the 'Bull'. Aldebaran or the 'red eye' of the bull is detached from the head and is not a part of the Hyades cluster but lies about a third of the distance between the Earth and the cluster itself. Since Aldebaran and the Hyades both lie on the same line of sight they appear related to each other. Lewis Boss realised that since the stars in the cluster were all moving towards the same vanishing point, they must be moving parallel to each other. The reason for them being the same age and in a cluster in the first place, is not so surprising when one realises that all the stars in the Hyades all formed from the same dust cloud or 'nebula'. Whilst they all have relatively small motions within the group, the whole cluster is whizzing through space at a great rate and all the stars in the group are moving parallel to each other. Lewis also determined that this vanishing point was located, just east of Betelgeuse - the big red star at the top left hand corner of the Orion constellation, but he went no further.

This is when Herztsprung had the bright idea of using the Hyades cluster of stars as his reference point to determine distances. Since the Hyades is close by in astronomical terms, in fact it is the nearest open cluster to us, over a period of many years we can detect the motion of the Hyades across the skies and it is this that enables us to use a sort of parallax to determine their distance.

This is where Art and Physics overlap. When an artist draws an outdoor scene, he includes 'vanishing points in the original sketch. A 'vanishing point is where all parallel lines appear to meet in order to give the drawing perspective i.e. to make the drawing appear three dimensional. So, when an artist wishes to draw a house, for example, he will first place two vanishing points on the drawing paper and draw the lines representing the parallel sides of the house as if they meet at one or other of the vanishing points.

Vanishing points are not new to astronomy as they are often given as the explanation as to why the Moon appears larger when it is rising or setting i.e. during the night, the Moon looks bigger when it is near to the horizon than it does when it is overhead. This is a problem that has puzzled astronomers and others for all time and written reports and supposed explanations go back to times beyond the ancient Greeks.

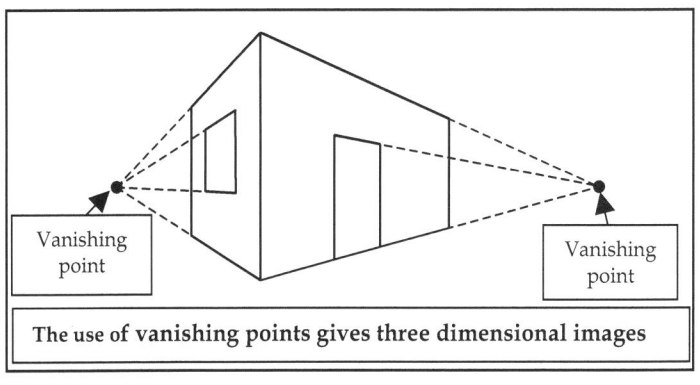

The use of **vanishing points gives three dimensional images**

The point is that whilst the Moon does 'look' much larger when it is close to the horizon than it does when it is high in the sky, in reality, it is actually smaller. The whole thing is an illusion. People often put the apparent change in size down to refraction or the bending of light as the light rays pass through the atmosphere but this is not the case at all. If anything, refraction makes the Moon appear a little smaller. One can measure the radius of the Moon's disc using telescopes or by the use of photography and this shows beyond any doubt that the whole thing is an illusion, the Moon stays the same size throughout the night. You can test this yourself if you do not believe me. The next time the Moon is rising and near to the horizon, looming large in the sky, either hang upside down from a tree or turn your back on the Moon, spread your legs apart, bend over and look at the Moon from between your legs. The Moon should now look its usual size and quite normal, the optical illusion is no more. You have to be careful that whilst you are looking at the Moon in this way, everyone else isn't viewing Uranus! To make matters worse, when the Moon is low in the sky, it is actually further away from us than it is when it is overhead, as you are a distance of about the radius of the Earth further away. This also has the effect of making the Moon look smaller rather than bigger.

So, just why does the Moon look larger when it is lower in the sky than when it is overhead (even though it is actually smaller)? The answer is, no body really knows! But, one suggestion is that it is to do with vanishing points. The closer an object is to a vanishing point, the bigger it looks. This is the basis of optical illusions and the Moon, on the horizon, is close to a vanishing point.

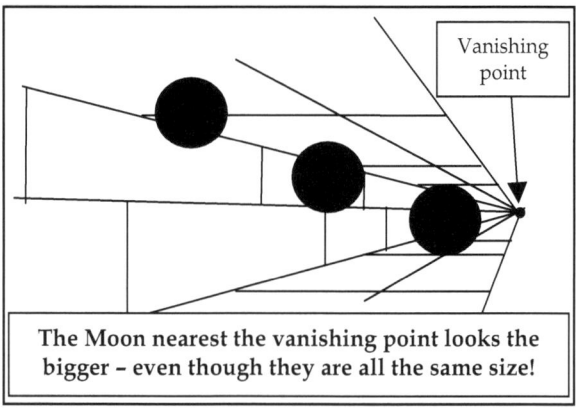

The Moon nearest the vanishing point looks the bigger – even though they are all the same size!

The three Moons in the diagram are the same size; measure them if you do not believe me, and yet the one nearest to the vanishing point appears much larger than the ones further away. However, if this were the case, then aircraft pilots, flying high in the sky should not see the illusion of a larger Moon when it is rising or setting because, to them, the Moon is not near a vanishing point. But they do; they too see the Moon as being larger at these times (unless they fly the plane backwards with their heads between their legs that is!).

Another explanation (and this is where the hot money is) is that when the Moon is near to the horizon, we can compare the Moon's size with other physical things, like buildings that we know the size of. We have a measuring stick against which we can compare the Moon's size. When it is near the horizon, we can see the Moon behind tall buildings or large mountains but when it is overhead we have nothing to compare its size against and so it looks small!

Does that mean that the illusion of the Moon's apparent change in size does not happen when one is out on a romantic cruise? That is, I hear you ask, when you are strolling the decks in the middle of the Caribbean, arm in arm with a delightful companion, extolling the virtues of a beautiful Moon, will it look bigger when it is close to the horizon, as here there are no sources of reference with which to compare the Moon against? There is nothing but sea in every direction. The answer is, no of course, the Moon still looks bigger! The illusion persists even when one has nothing with which to compare the size of the Moon. It still appears larger when it is near to the horizon in the middle of the sea. This brings me back to my original answer. The explanation of the illusion of the apparent change in size of the Moon (and illusion it most definitely and scientifically is) is that we just don't know! On that we will return to the problem of finding the distance to the Hyades cluster using vanishing points. The stars in the Hyades open cluster are all moving together in a parallel direction and that is the direction in which the nebula or gas cloud in which the stars originally formed is moving relative to the Earth. The Hyades are close to us in astronomical terms and so we can measure their actual change in position over a short period of time and hence determine their motion. Since they are all moving parallel to one another (we can ignore their actual movement within the cluster as this is small compared to the motion of the cluster itself) they will appear to us to be moving towards their 'vanishing point'. Hence the vanishing point can be found by representing the motion of each star with a little arrow. Extending these little

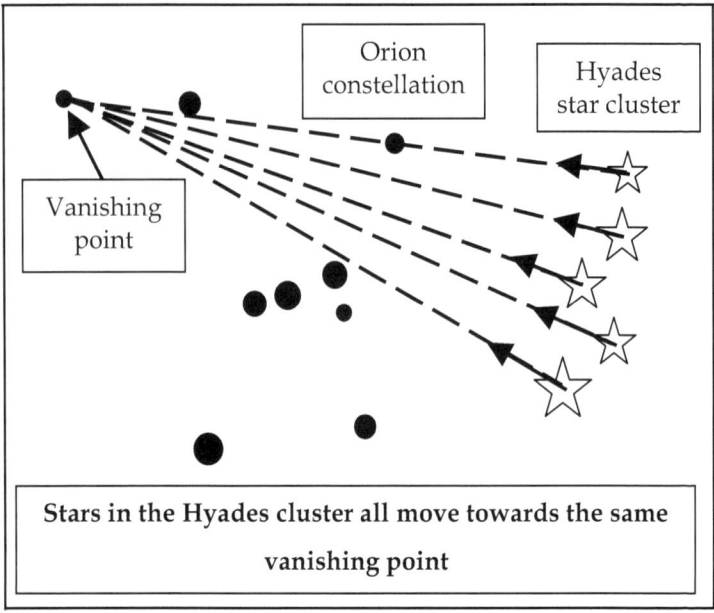

Stars in the Hyades cluster all move towards the same vanishing point

arrows, we find that they all meet in the same place and this is the vanishing point, which is just east of Betelgeuse - the big red star at the top left hand corner of the Orion constellation. This was the discovery made by Lewis Boss at the Dudley Observatory.

If we now measure the angle between this vanishing point and the Hyades cluster we find that it is thirty three degrees and so 33^0 is also the angle between the direction of motion of the cluster and the 'line of sight' between the Earth and the cluster. The reason being, all parallel lines meet at the vanishing point. If we look at the vanishing point too, then we must be looking in a direction parallel to the direction in which the cluster is moving. So if we draw an imaginary line between the cluster and us and look at the vanishing point, the angle between this imaginary line and the direction in which we are looking is the same angle as that between this same line and the direction in which the Hyades cluster is moving. Hence, we now know the direction in which the Hyades cluster is moving – at 33^0 to the line of sight of the Hyades.

The next step in determining the distance to the cluster is to measure the 'radial' velocity using the Doppler Effect. The 'radial' velocity is the stars velocity towards or away from us along the 'line of sight'. We will look at the Doppler Effect in great detail in a later chapter but briefly; it works in the same way as radar speed traps. If a source of waves is moving away from us then light emitted from the source is stretched. If the source is moving towards us then the light emitted by the source is squashed. The faster the source moves the more the light is stretched or squashed. This is how we determine the speed of your car and also how we can measure the 'radial' velocity of a star - by seeing how much the light arriving from the star is stretched or squashed. We find that the Hyades cluster is moving away from us at a speed of just over 40 km/s.

Knowing the radial velocity and the direction in which the cluster is moving we can work out the 'transverse' velocity of the cluster. This is their velocity across, or at ninety

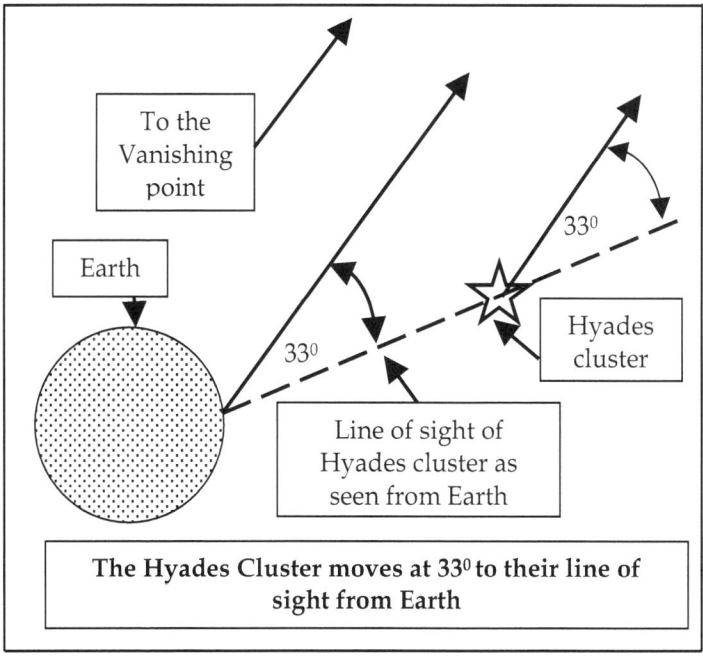

degrees, to the line of sight. A bit of simple trigonometry tells us that the transverse velocity is given by:

$$\text{Transverse velocity} = 40 \times \tan 33^0$$
$$\text{Transverse velocity} = 26 \text{ km/s.}$$

This is the velocity of the Hyades cluster across our line of sight.

Using the formula, distance travelled equals the speed multiplied by the time taken we can work out the actual distance travelled by the Hyades cluster, across our line of sight, during a period of ten years.

However, we must convert the time of ten years into seconds first. The number of seconds in ten years is 3.2×10^8, and so our calculation becomes:

$$\text{Distance travelled} = (\text{speed}) \times (\text{time})$$

Distance travelled across our line of sight = $26 \times 3.2 \times 10^8$

This tells us that the Hyades cluster moves a distance of 8.2×10^9 km across our line of sight during these ten years. That is, over five times the distance from here to the Sun every year. Quite a way isn't it?

From Earth we can only measure the angle through which the Hyades have moved over the same ten year period and the angle through which they move is found to be 0.00033 degrees. To give some idea as to how small this angular movement is, we can compare it to the Moon. The Moon subtends an angle of about half a degree as seen from the Earth. Consequently, it will take nearly sixteen thousand years for the Hyades to change in position by one Moon diameter. Knowing the distance the cluster has moved across the line of sight and the angle the cluster has moved through during this same time, gives us the distance to the cluster

Distance to cluster is equal to distance moved across line of sight divided by the tangent of 0.0033^0.

Distance to Hyades = $8.2 \times 10^9 \div (\tan 0.00033)$

This gives the distance to the Hyades cluster as 1.4×10^{15} km or 150 light years.

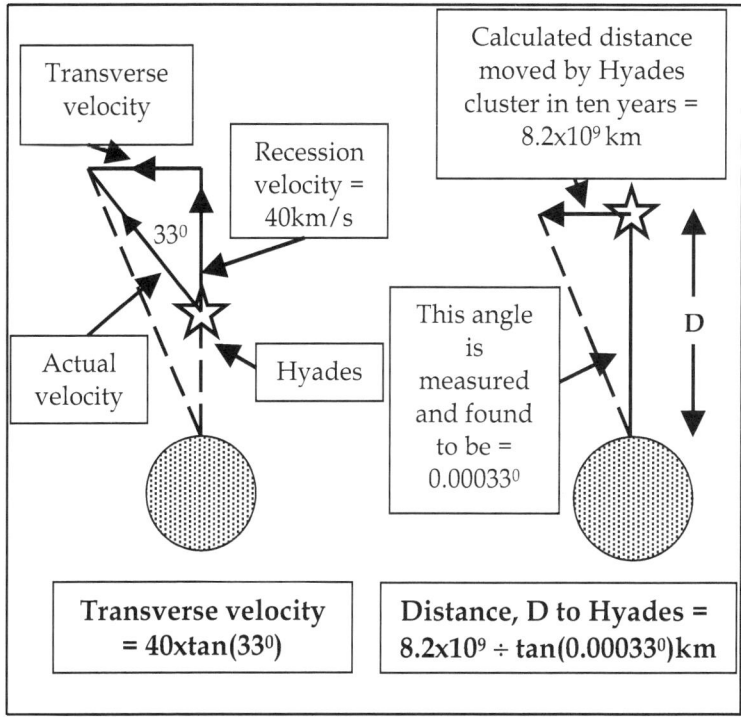

From this, we now know the distance to the nearest star cluster, which is certainly interesting, but unfortunately, there are no Cepheid variables in the Hyades cluster! Star clusters tend to be young so that the individual stars motion, or 'peculiar' motion has yet to send them on their own way away from the group. The stars are still in the nursery and have yet to flee the nest. Cepheid variables are old stars going through their death throes. However we now have an accurate method to find the distance to other star clusters that are too far away to use this method of statistical parallax. What we do first is to construct a Hertzsprung Russell Diagram for the stars in the Hyades cluster. We then look at our next star cluster and draw up a Hertzsprung Russell Diagram for the stars in that cluster. These stars will have apparent brightness of less than those of the nearer Hyades cluster because they are further away. The 'reversed seven' pattern on the diagram will be lower and lower the further away the star cluster is. We then see just how much we have to move the HR Diagram up until it 'fits' the HR Diagram for the Hyades cluster. By using the 'inverse square' law, we can then tell how much further they are away from the Hyades cluster – and we know how far away that is now. Measuring distances by HR 'fitting' is far more accurate than using the colour of a single star to determine distance, as now we are dealing with hundreds of stars and any variation from star to star cancels out when we use such a large sample.

Hertzsprung found several star clusters containing one or two Cepheid variables in each. He measured the distance to the cluster by 'Hertzsprung Russell fitting' and then found the actual brightness and period of the Cepheid in each cluster. In this way he was able to calibrate the Cepheid distance scale. Hertzsprung could now measure the period of the Cepheid and determine its actual brightness. He measured its apparent brightness here on Earth and from this he could determine just how far away the Cepheid was. He now had a way to measure intergalactic distances and using this method he became the first to measure the actual distance to the Small Magellan Cloud, and got it totally wrong. The actual distance to the Small Magellan Cloud is fifty times larger than the one determined by Hertzsprung. This is so far out that some put the discrepancy down to a misprint in the paper or a miscalculation on Hertzsprung's part. It is true that Hertzsprung assumed that the intervening space between the Earth and the clusters was as 'clear as a bell' and we now know that this is not the case. Dust in space 'dims' distant stars but this alone does not account for Hertzsprung's error. However, methods were refined and the Cepheid distance scale became an important tool for measuring distances in space.

Did you see the film, "Good Will Hunting"? This is where the janitor in a University solves mathematical problems, which the lecturers had left behind on the board in the lecture theatre. Well, our next character is something along these same lines.

In 1905 Milton Humason dropped out of school at the age of fourteen and took up a job at the Mount Wilson Hotel, as a general dogsbody with the responsibility of looking after the mules. By 1910 he had come up in the world and become a mule driver, driving the mules with their heavy loads of equipment up to the top of Mount Wilson where the new 100-inch reflecting telescope was to be built. It was whilst doing this job that he met and married the daughter of the observatory's engineer. With the added responsibility of a family, he took the post of head gardener on a Pasadena estate before purchasing his own citrus ranch.

When a vacancy came up at the Mount Wilson observatory, his father-in-law told him to put his application in. This he did, with the result that he left the citrus farm (which had turned sour anyway) and took the job. Whilst performing his duties, he became friendly with the students and astronomers alike and gradually learned how to be an astronomer - being awarded the post of 'assistant astronomer' in 1922. During his time as night assistant, Humason had been given some photographs of the Andromeda galaxy to study by the astronomer, Harlow Shapley. Humason thought he could see some Cepheid variables within Andromeda and marked them in ink on the photographic plate so that he could show the variables to Shapley the next day.

Shapley was a firm believer that our galaxy, the Milky Way, was all that there is and made up the entire Universe. Because of this he believed that Andromeda was just a gas cloud within the Milky Way. He had just had a 'great debate' with Heber Curtis on this matter as Curtis believed that the Milky Way galaxy was, in astronomical terms, quite small and was just one of many such galaxies in the Universe. Shapley rejected Humason's suggestion, explained the reasons for his beliefs to Humason, and then told Humason to mind his own business. He cleaned the identifying marks from the photographic plates and went off in blissful ignorance - leaving the discovery for others to find.

Humason, whilst being of lowly scientific origins, was a great and respected observer and eventually teamed up with the great man himself, Edwin Hubble. Hubble too found Cepheid variables in both the Andromeda and the Triangulum galaxies but instead of ignoring them as Shapely had done, Hubble looked further. Firstly he measured the average apparent brightness and the period. He found that they agreed with the law developed by Leavitt but since Hertzsprung had now calibrated the Cepheid variables, he could actually calculate their true brightness and hence determine how far away they were. He found that the Cepheids in these galaxies were about 930 000 light-years away and far more distant than anything imagined before. Without doubt, this settled the question of the size of the Universe. The size of the Milky Way was known and it was far smaller than the distances to the Andromeda and Triangulum galaxies and so these galaxies must be well outside our own galaxy. Had Shapley listened to Humason, perhaps he could have made this discovery but he did not so we will never know! We will meet Hubble later in our story - but suffice it for now to say that Hubble had just determined the scale of the Universe, a great discovery in itself.

With the calibration of the distance ladder, and Cepheid variables in particular, we can measure the distance to far off galaxies. Hubble's determination of the distance to the Andromeda galaxy showed that Curtis had been correct in his arguments in 'the Great Debate' in saying that the Universe consisted of billions of other galaxies apart from our own. Who won, in the 'Great Debate' I hear you ask? It doesn't matter, scientific fact is not proven by debates, it is proven by experimental results, though Curtis, who was in the right, is said to have blown it!

As for us, this marks the half way stage in our journey to the Hubble constant since we are now able to measure distances accurately to most galaxies. We will now go on to the second part of our journey, which will enable us to determine the 'velocity' at which they move, but to do this we must ask the question, "What is light"?

Chapter 5. Strange stuff is light.

Unless we are looking straight at something we don't see it. If someone is talking about us behind our backs we can still hear them even though our ears are pointing outwards. Even though our nose is pointing forwards, if someone in the room makes a rude smell we can still smell it, eventually, as the aroma diffuses all around the room. However, if someone makes a rude visual sign behind our backs then we don't know it has happened until someone else tells us about it. We have to be looking straight at them otherwise we don't see it. The other thing that is different about vision is just how easy it is to switch sight on and off. Close your eyes and everything around you disappears, just as if everything has gone away.

I remember several boring conversations in my fresher year at University. After vast quantities of ale to stimulate the thinking process or, more likely, to arouse the more argumentative side of my nature (doesn't take much to do that actually), Philosophy students would try to convince us 'freshers' that when we closed our eyes everything else disappeared. That is, everything is a figment of our imagination. On protesting that this was utter rubbish, we would be required to provide evidence that this was not the case. That is, prove that everything around us actually does exist and that we are not just imagining life as if it was one big dream – one cannot do this of course. However, it doesn't work the other way around. If you tell someone that when you close your eyes they disappear because they are a figment of your imagination then they know full well that this is not the case. They happily continue to drink until all the beer has gone whilst you sit there with your eyes closed trying to tell them that they don't exist anymore.

But somehow, light is like that. Unless we point our head at something we don't see it. Unless we choose to open our eyes we just don't see it. It is like our own personal view on the world. It is little wonder that the Greeks had such strange ideas as to how light behaved. One idea they had was that objects shed an outer layer in the same way as a snake sheds its skin. These layers would travel towards the observers' eyes getting smaller and smaller as they travelled until they could enter the eye and the image was received. As to what happened when skin after skin from an object arrived in the eye and eventually filled the eye up, we don't know, as they made no attempt to explain how the eye itself worked. Perhaps this is where the phrase 'getting an eye full' originated!

Later Greek theories had the eyes made up of the four basic elements, earth, fire, air and water, four elements that they believed made up just about everything in one way or another. However, it was Aphrodite herself, the Greek Goddess of love and lovers, who is thought to have put the 'fire in the eyes'. She had had quite an eye watering start herself. Her father Uranus had been in a fight with Cronus (a wonderful person who married his sister and ate his own children as soon as they were born). Uranus lost the fight and ended up being castrated. His genitals, which fell off and sank into the sea were his 'seed' and somehow managed to produce Aphrodite. She emerged from the foaming sea some nine months later and, as we now know, put the fire into everyone's eyes. This is why Greek statues had jewels in the eye sockets to represent the 'fire'. Taking this 'highly scientific idea' one step further the Greeks decided that, if this was the case, particles must stream out of everyone's eyes, bounce off objects and return to give us sight. It didn't seem to bother them that this would also mean that they would be able to see in the dark! There were one or two dissenters who complained about this result and suggested that the rays emitted from the eyes somehow interacted with different rays emitted by the light source to produce vision. This would overcome the problem of us not being able to see in the dark but these ideas were largely ignored - on the basis that after all, animals seemed able to see in the dark even if we couldn't!

So to all intents and purposes, for the next 1000 years or so people went around convinced that their vision came from a beam emitted from their own eyes. In order to see they thought that they had to a) point their head and eyes directly at the subject, b) open their eyelids, c) a beam of particles would stream out, illuminate the object and return the information to their eyes. Whilst they could not see at night, this was not a problem because cats etc could all see in the dark and so the theory must be correct.

Round about 1000 AD Ibn al-Haytham a scientist from the city of Basra in Iraq eventually saw the light. He reasoned that because you could have 'pin hole' cameras or 'camerae obscurae' then light must be a beam emitted by the light source – 'camera' in Latin has the meaning 'room' and 'camera obscura' literally means 'darkened room'. Because an image could be produced on the wall of a darkened room having a small hole in the window blinds opposite, he argued that anything emanating from the eye was redundant and so was not needed. He also reasoned that not only was vision an effect caused by beams emitted by a light source but also that these beams did not travel 'instantly' but had a velocity and so light took a certain time to reach us. Since his work would not be on 'general release' in Europe for a good six hundred years or so, Europeans were left fumbling in the dark for some time to come. Whilst there would be several advances such as how the camera obscura worked, how the eye worked with its upside down image, and the inverse square law (virtually all these advances being due to Kepler at the turn of the seventeenth century) no advances had been made in the nature of light. That is, just what is light? They knew that it was something given out by a light source, travelled at a finite speed, travelled in straight lines, refracted and so on but they did not know just what this 'something' was. For advances in that field we have to wait for Newton, Hooke and Huygens.

When one discusses the work of Hooke and Newton it is difficult to separate the pair as their paths and theories are intertwined and overlap and this often resulted in disputes over the 'priority' of ideas. The lack of scientific journals in which to publish their theories and experimental results meant that scientists had to correspond with each other by letter and this caused problems when it later came to deciding whose idea it was originally. One way of claiming priority to an idea was to send the result in the form of an anagram and leave it to the recipient to decipher what had been discovered. Of course, several of these anagrams were deciphered wrongly and the originator could be credited with a discovery that he had never even studied, leaving his actual discovery open for others to claim as their own.

Robert Hooke sprang into life on the Isle of Wight, UK, in 1635. The son of the local vicar, he went from one illness to another and so he had to be educated at home by his father until he was thirteen years old when, unfortunately for Robert, his father passed away. At this point the young Hooke, having been left the grand sum of forty pounds, was sent to London where he took up an apprenticeship as an artist. He then moved on to Westminster School before leaving to attend Oxford University and, if only for the money, become a chorister at Christ Church Oxford. Whilst Hooke never graduated, Oxford was seething with scientific thought and Hooke would have met some of England's leading scientists during his time there - resulting in his appointment as assistant to Boyle (as in Boyle's law of gases). Science was largely the pastime of rich aristocrats who would 'play' at science for something to do. The 'Royal Society' had been formed by these well to do posers and, quite naturally, they wanted entertaining. They required somebody to come up with a few different but interesting experiments or ideas that they could then discuss at their weekly meetings. Hooke got the job and became 'Curator of Experiments' at the Royal Society. One of the items Hooke developed to amuse the members was the compound microscope with its accompanying system of illumination. With this, and his experiences as an artist's apprentice gained before he went to Westminster, he was able to make accurate drawings of fleas and the cell structure in samples of cork. In fact it was Hooke who gave Biology the term 'cell' as the regular rectangular cell walls reminded Hooke of the bare rooms or 'cells' used by monks in monasteries. Hooke spent a great deal of his time studying astronomy and, in order to show that the Earth was moving, even knocked holes through the floors and roof of his house so that an enormous vertical telescope could be installed. What he was looking for was parallax in the stars - as this would provide direct evidence that the Earth was moving. The telescope had to be vertical to prevent the refraction of light by the Earth's atmosphere affecting his results. If he had persevered with his measurements he may well have demonstrated the movement of the Earth but the pressure of having to produce new idea after new idea, week in week out, year in year out to pander to the members of the Royal Society meant that there just wasn't the time for lengthy experiments. After all, he needed the job, as he wasn't a rich man. Apart from ancient astronomers, who had been paid by various Kings to study the stars in order to compile horoscopes that could be used to give advice and therefore help the King to rule the country (as well as to tell them when the best time was to invade an adjoining country), Hooke was probably one of the first professional Scientists.

Time was always the problem, and many of his experiments went on to be developed by others who took the credit for them (it must also be said that Hooke did not have the mathematical skills to prove his ideas analytically). This situation led to a great deal of acrimony and accusations of plagiarism between Hooke and some other scientists of the day and this was made even more unpleasant as several of the combatants were members of the same Royal Society. Hooke had a momentous 'difference of opinion' with the Dutch scientist Christian Huygens over who had discovered the spring escapement in clocks. This paled into insignificance when compared to Hooke's running battle with Newton over the theory of gravity and optics.

Amongst his other claims to fame, Hooke was also an architect. He was instrumental along with Wren (a long time friend of Hookes) in the rebuilding of London after the great fire of 1666. Whilst Hooke largely concerned himself with boundary disputes he did design a number of buildings including the new 'Bethlehem Hospital'. This was the World's first lunatic asylum, better known as 'Bedlam', where anyone could pay one penny to peep through the door and be amused by the inmates hanging naked from the wall. For a further fee the 'nurses' would excite the patients until they went berserk - with all the profits going to hospital funds to improve the amenities of the patients. It must be said that Hooke was responsible only for the design of the building and not the medical treatment meted out to patients. Hooke, like any good scientist kept a record of what he did including a diary that was meticulously completed throughout his life. It contained what he did, what he ate, whether the food agreed with his digestion or not, it listed the nights when he slept with his female servants and he used a code to denote whether he had climaxed or not. His niece lived with him for some time and it is thought that she was one of the few ladies that Hooke truly loved, which is perhaps as well, because she has many coded entries in the diary! Her father James, Robert's brother, came to owe Hooke a great deal of money and so could say little even if he wanted to. James took his own life by hanging himself with the consequence that the local authorities seized James' assets as a punishment for the crime (which was a bit late for that really since he was now dead), and left the rest of the family destitute!

Newton was born on December the 25th, 1642 (the same day that Galileo died) in Woolsthorpe, Lincolnshire; Christmas day except, since they did not use the Gregorian calendar then, it would be January 4th 1643 on the calendar we use now. His father died three months before his birth and, when Newton's mother married the local vicar two years later, he was fostered out to his grandmother who brought him up. Despite his school reports describing the young Newton as 'idle' and 'inattentive' he finally went to Trinity College, Cambridge University and graduated in 1665. It was at this point that the University had to be closed due to the plague and Newton returned to his home in Lincolnshire, spending the next two years inventing calculus and developing his theories on light and gravity. When the University reopened two years later, Newton at the age of 25, having lived through a civil war, the plague and the great fire of London, returned to Cambridge to take up a fellowship. Newton is sometimes remembered as the greatest of all scientists, one of the first scientists to use the 'scientific method' where theories had to be backed up by experiment.

In the 1930's it emerged that Newton had a darker side to his life and as it turns out, a very much darker side to his life. Two of his friends who were present at his deathbed became the first 'spin doctors' and swept the darker side of Newton's life 'under the carpet', labelling documents describing Newton's activities as 'not fit to be printed'. Had these papers come to light during Newton's lifetime, they could have led to Newton being thrown out of University or even hung from the scaffold. The friends paid for Newton's extravagant and expensive monument in Westminster abbey, extolled Newton's virtues and wrote his first biography – neglecting to mention his darker side.

In 1936, Newton's private papers came up for auction at Sotheby's and a famed economist, John Maynard Keynes, bought one hundred of the lots. Newton, as was the custom with all leading scientists at the time wrote his notes in code to stop anyone else reading them and stealing his ideas and it took Keynes six years to decode the private manuscripts. When he had finally completed the task he was appalled, as up till then people had only known of Newton as a great scientist and none had known of his desires to be a 'magician'.

Newton was a pious, religious fanatic and in his undergraduate years had been disgusted with his fellow students who lived the life of 'wine, women and song'. Newton thought that the only way to resist these temptations was to devote himself to his studies and moved in with a like-minded student, one John Wickins. Whilst Wickins was socially 'higher' than Newton he eventually became Newton's secretary and they shared the same rooms for the next twenty years, the arrangement only coming to an end when Wickins gave up his fellowship in order to marry. Was there a relationship between Newton and Wickins? Who knows or cares, but the abruptness with which the split occurs leads some to say that there was.

Students at Cambridge were required to study the teachings of Aristotle but Newton also read the philosophy of Rene' Descartes who put forward the idea that God had made the Universe like some giant clockwork machine and, once he had it all up and running, he left the Universe to its own devises and had nothing more to do with it, never oiling it even once. Newton felt that this could not be the case and believed fervently that God had a hand in everything that went on in every minute of every day and included this in his writings. References to God were removed after his death to leave the cold mechanical laws of gravity behind the motions of the Heavens.

Newton had two well-kept secrets, his alchemy and his being a heretic. Both of these addictions were driven by his strong religious beliefs. On returning to Cambridge after its reopening, Newton equipped his rooms with everything necessary to smelt and mix metals of all descriptions in the pursuit of the 'philosopher's stone'. This is the catalyst that will turn base metals into gold and was the goal of alchemists from time immemorial. Newton believed that when God made the Universe he left the information of how it was done with his disciples on Earth. These included Noah and so Newton believed that the information was fragmented but could be found as coded recipes in the Bible, ancient texts and Greek mythology. Newton did not want to make money out of it or to use this knowledge to perform magical feats; he just wanted to imitate the works of God as part of his scientific enquiry.

Unfortunately for Newton, this was not the case with other alchemists. They had been defrauding people out of great amounts of money by the promise of turning base metal into gold, so much so that alchemy had been declared illegal. The punishment was hanging from a gilded gallows and, to enhance the spectacle for the observers, the doomed alchemist would be covered in metallic foil before being strung from the gallows. For this reason, Newton was extremely secretive about his activities and did not let even his assistant know what he was looking for. Whilst Newton secretly became the leading alchemist in Europe, alchemy would remain the one area where Newton had tried and failed.

On the religious side, Newton believed that both the Anglican and Catholic churches had defiled Christianity by preaching the Holy Trinity – the father, Son and Holy Ghost. Newton felt that this went against the first commandment – there is but one God and so he refused to believe in the Holy Trinity. This was bad enough for any normal person who, if discovered, would be branded a heretic and could be jailed or set upon by the mobs. These were gangs of people, encouraged by church ministers, who would set upon, kick, beat or hang anyone described as a 'heretic'. In Newton's case it was even more complicated as he was a professor at Trinity College, a college set up by Henry the Eighth dedicated to the doctrine of 'Trinity'. If discovered, Newton would surely be thrown out of Cambridge. As it was, as a professor at Trinity he was required to take up Holy Orders. Newton managed to obtain a special dispensation from Charles II to exempt him from this.

But, Newton kept his secrets until his deathbed and left it to his friends to cover them up. Whilst he had been pursuing his alchemy his discoveries made during the plague years drove him to promotion and he became Lucasian Professor and lectured on optics. Newton hated the students and they had an equal dislike of him. As a consequence, few turned up to his lectures and of those that did, hardly any understood what he was teaching them.

It was the theory of optics that caused Hooke and Newton to have their first clash. The experimental result that colours could be obtained from white light had been known for a long, long time. Rainbows appearing after rainstorms 'a sign from God that the storm is over' as we are all told by our Mum's when we are children is a good example of this. We also get rainbows whenever we have faulty optics. You don't get much of this these days, but when you look through a toy telescope with plastic lenses objects tend to have 'rainbows' around the edges. We now know that this is caused by a phenomenon called 'chromatic aberration' and this is a similar effect to the rainbows found after a storm. Light travelling through a poorly cast piece of glass used in windows in olden times would provide the same effect. It was also known that a spectrum could be obtained by using a feather. Darken a room by closing the blinds having first made a small hole in them and hold a very fine white feather next to the hole. A spectrum could be formed on a piece of paper or on the wall opposite. This feather was the first kind of diffraction grating. From the times of Aristotle, people had believed that white light was 'pure light' and that modifying or corrupting this 'pure' white light produced the colours.

Hooke had published his book 'Micrographia' in 1665, which contained drawings of fleas, cork, insects, feathers and a variety of 'minute bodies' made visible with his compound microscope. At the back of this book, Hooke had included an addendum containing various theories and results from a wide range of scientific enquiries having nothing to do with microscopy. One of these theories was on light.

Using his compound microscope, Hooke had investigated the effect of the thickness of mica on white light. He had shown that the thickness of the mica did affect the colour of the transmitted light and thus his results supported the theory that pure white light was 'corrupted' as it passed through the mica and became coloured. He had also found the results to be 'repeatable' – a requirement in any scientific theory, but he could not establish any pattern. In this same publication Hooke also suggested that light was a wave, like the ripples on water and also put forward the view that the waves were transverse – that the 'particles' vibrated in a direction perpendicular to that in which the wave travelled.

At this time it was known that one could produce an excellent rainbow, or 'spectrum' as Newton called it (spectrum being a Latin word meaning 'image in the soul'), by passing the light from the Sun through a triangular glass prism. A prism is a shape that has the same cross-section all the way along its length and so the prism used here is something like a Toblerone bar, no matter where you cut it along its length you get the shape of an equilateral triangle.

Prior to Newton they explained the spectrum formed by a glass prism as being due to light following different paths through the prism. Red light only travelled through the thinner, uppermost part of the prism and so it travelled through less glass and was therefore deviated the least. Violet light, being deviated the most travelled through a thicker part of the prism and hence travelled through more glass whilst inside the prism. The conclusion that they jumped to was that the light was changing as it went through the glass – hence Hooke's attempts to verify this theory with his pieces of mica. It had been thought that the light had been corrupted as it passed through the prism.

It had started off as white or 'pure light' and gradually changed to red then to orange as it travelled through more and more glass until eventually it ended up as violet when it had passed through a sufficiently big 'chunk' of glass. It appears that no one thought to check the interpretation by passing the red light through another piece of glass to see if it changed its colour to orange; until Newton came along that is.

Instead of thin sheets of mica, Newton investigated the light travelling through a prism. In 1666 at Cambridge, Newton darkened the room and allowed light from the Sun to pass through a circular hole and fall onto a prism. He then took the resulting spectrum and separated the red light by allowing the spectrum to fall onto a card containing a slit where the red light appeared. Using the red light passing through the slit, he passed the red light through another piece of glass, reflected it and did everything imaginable with this ray; but whatever he did, it remained red. Furthermore, by using a second prism, he found that he could recombine the colours of the spectrum formed by the first prism back into white light. This led Newton to develop the theory of colours in light that we use today – that white light is not a colour but a mixture of all the colours of the rainbow. Rainbows themselves are caused by white light being refracted and dispersed by droplets of water left in the atmosphere by the rainstorm.

When the light from the Sun, or 'white light' as it is termed, passes through the prism, it is 'bent' or deviated from its original path and this results in the white light being split up, or 'dispersed,' into its different component colours. The colours are always in the same order Red, Orange, Yellow, Green, Blue, Indigo and Violet and they appear in the same order as those in a rainbow after a storm (In Physics though, there is no 'pot of gold' at the end!). We now know that white light is not a colour in itself but a mixture of all the colours that our eyes are sensitive to. In a vacuum, or air for that matter all the colours of light travel at the same speed (300,000,000 m/s) but in a medium such as glass they all travel at different speeds. Red light travels the fastest in glass whilst violet travels the slowest.

Imagine that whilst you are driving along the road you inadvertently hit a lamppost with the 'nearside' wing of your car. This wing is slowed down but the 'off side' of the car is still going fast and so what happens is that the car changes direction. It swerves around and mounts the pavement.

Had you crashed 'head on' into a wall this would not have happened, both sides of the car would be slowed down together and the car would be brought to a halt in a straight line. Hopefully, in all this risking of life and limb in the interests of science you will come out of this in one piece and unchanged.

It is the same with light. If a beam of light strikes a glass surface 'head on,' then both sides of the beam of light slow down together and the beam continues in a straight line and is not deviated. If the beam of light strikes the surface of the glass at a glancing angle then one side of the beam slows down before the other and the beam is deviated from its original direction, it swerves with the side that was slowed down first on the inside.

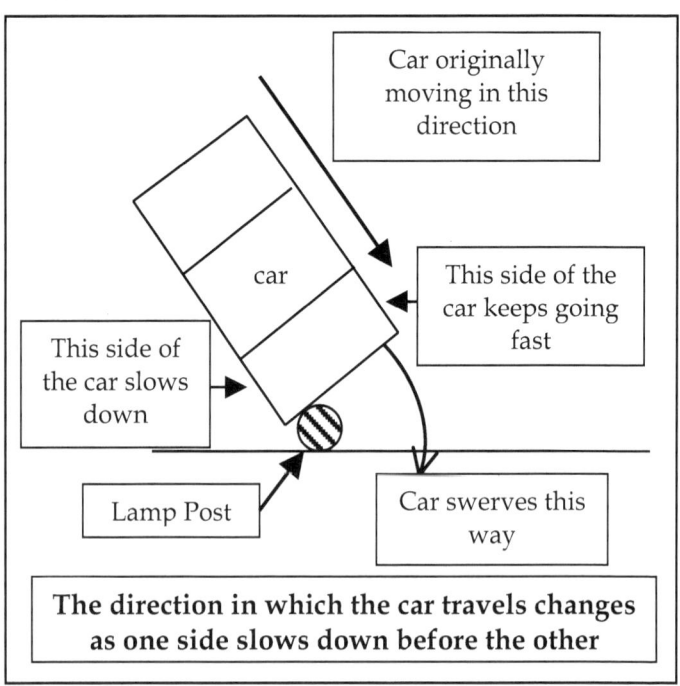

The slower the beam of light travels in the glass then the more it is deviated. In glass, of all the colours, red light travels the fastest and so is 'bent' or deviated the least from its original path. Violet light is slowed down the most by the glass and so is deviated the most.

The Greeks and more recently Rene' Descartes had proposed that light consisted of a stream of particles that hit the eye and Newton developed this theory and called the particles 'corpuscles'. He proposed that what the prism did was to filter these corpuscles out and separate them into their various types or colours. White light was not 'pure light' but a mixture of all the various particles. Newton published a paper on his results in the 'Philosophical Transactions of the Royal Society' – the first scientific Journal. Whilst Newton's views were respected by most, Hooke and Huygens literally pulled it to pieces. They both

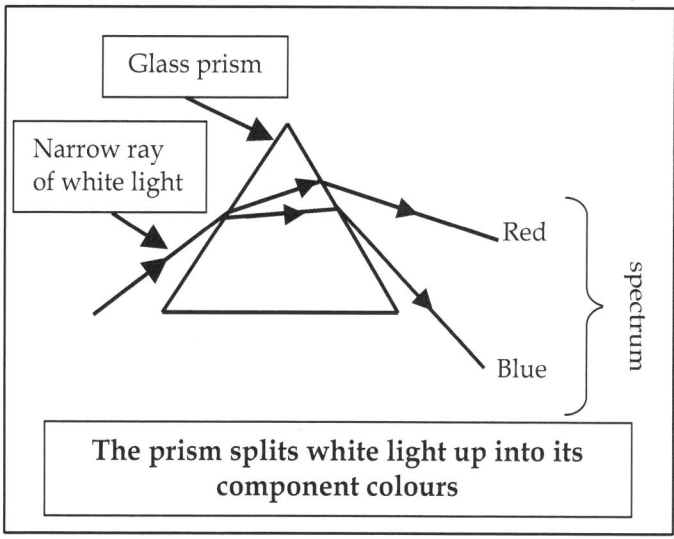

believed that light was a wave and they criticised the way Newton had relied solely on experimental reasoning rather than including some philosophical arguments! Hooke also felt that some of the ideas had been stolen from him, and this resulted in some acrimonious comments from Hooke - such as Hooke's critique of Newton's work on optics where he said that what was original was wrong and what was right had been stolen from him!

We may have spent a little too long discussing Hooke and Newton but this argument over whether light is a wave or a particle has continued right up to modern times. Newton's corpuscular theory won the day and was the generally accepted theory on light throughout Newton's lifetime. Had Hooke not had a vitriolic dispute with Huygens over the watch escapement the two might have teamed up and jointly promoted the wave theory of light but as it was, divided they fell and Newton won the day.

Perhaps we should say something about Huygens who also proposed that light was a wave (if only because, at the time, he did most of the work on this theory). Christiaan Huygens was a Dutch mathematician and scientist born in The Hague in 1629. Amongst other triumphs he did the mathematical calculations on the pendulum and patented the first pendulum timepiece with a view to solving the longitude problem. He derived the mathematical solutions to circular motion, which allowed Hooke, Newton and others to work out planetary motion. He also designed and built the finest telescopes of the day, which enabled him to discover the rings around Saturn and one of its moons. Huygens also put forward theories supporting the wave theory of light. Huygens met Hooke and Newton on several occasions and was himself elected to the Royal Society.

Since Hooke and Newton were now both 'big men' in the Royal Society, they could hardly feud forever and so they decided to correspond by letter so that there could be no misunderstandings. They managed three such letters before the troubles began again. In Hooke's third letter, he had written to Newton proposing the principle of inertia, the idea that a body will keep on moving unless a force acts on it to stop it. He also proposed to Newton the idea that the gravitational force between two Heavenly bodies was the same as that on Earth, dependent upon their respective masses and followed an inverse square law (move two bodies twice as far apart and the gravitational force between them reduces to one quarter). That is as far as it went. He didn't derive the formula mathematically as

Newton did, partly because Hooke didn't have the mathematical background and partly because Newton, who had invented calculus, kept this mathematical tool to himself to give him 'an edge' over other scientists. Hence the row between Newton and Gottfried Leibnitz over who had discovered calculus as Liebnitz went on to develop it independently and claim it as his own, not knowing that Newton had done this earlier and kept it a secret. Nor did Hooke apply his proposed theory of gravity, as Newton did, to calculate the time for the Moon to orbit the Earth and thus check the theory experimentally.

Newton never replied to this letter from Hooke (did he receive it?) and always insisted that he had formulated his ideas on gravity on his own. When Newton published 'Principia' he refused to include a mention to Hooke anywhere within the book, not even in the preface. Hooke, who had been a socialite with several good friends, became an embittered old man. The two remained enemies and Newton delayed publication of his complete works on optics, the book 'Opticks' until after Hooke's death, no doubt to stop any criticism from that quarter. Hooke was 'vertically challenged' i.e. not very tall and a medical condition had caused him to become crooked. Since Newton had referred to Hooke at least once as a 'dwarf' it is quite certain that when Newton's is quoted as saying that If he had seen further than anyone else then it is only because he had stood on the shoulders of giants, he is passing an insult at Hooke - stating that he had not stolen any ideas from a little runt like him!

One wonders what Hooke would have done if he had known of Newton's two secrets, alchemy and heresy. Would he have reported Newton to the authorities? We will never know.

There was a portrait of Hooke hanging in the Royal Society but this disappeared long ago with the result that, apart from written descriptions, no one knows what Hooke looked like! It is often alleged that on Hooke's death Newton, who could still break into a fit of temper on the mention of Hooke's name twenty years after Hooke died, took the portrait and burned it.

In true Hooke style the story of his portrait is still embroiled in dispute. On the three hundredth anniversary of Hooke's death (2003), there was a great deal of interest in Hooke with the end result that two portraits have been found, both claiming to be that of Robert Hooke. In the red corner is the portrait that was found in the Natural History Museum by historian Lisa Jardine and appears on the cover of her biography on Hooke. It was apparently lurking in the Museum with the name of John Ray on the portrait (did Newton switch name plates on the picture?) but the blue corner say the picture bears a resemblance to other portraits of John Ray – a seventeenth century botanist and so it stands a good chance of being that of John Ray. Meanwhile back on the Isle of Wight in the blue corner, a local historian has found a mortgage deed signed by Robert Hooke and bearing his seal. The seal has a picture on it and it is claimed to be a likeness of Hooke. The red corner rejects this claim on the basis that people at this time just didn't have seals with their own picture on it. But of course, this is Robert Hooke we are talking about, he was first with just about everything else so why not a seal with his picture on it? Needless to say the two pictures do not bear any resemblance to each other!

This time period is important to us in our road to the Big Bang as this was the first time that light was known to consist of various colours. This presented a major leap forward to Scientists who now had another tool with which to prise open the secrets of the Heavens. It also marks the beginning of the controversy over whether light is a wave or a particle. William Herschel made the next leap forward on our journey.

William Herschel was a musician in the German army and, during the battle of Hastenback in the seven years war, which started in 1756, he found himself cowering in a ditch hiding from the French army. It was at this point that Herschel decided that a life in the army was not for him and, when the French invaded Hanover, he escaped to England. This was not a random choice as he had visited England before when his regiment, the Hanoverian Foot Guards, had been stationed there. Whilst in England Herschel had learned English and formed friendships with several English musicians. In 1766, he eventually settled in the English city of Bath (the same city where Pigott eventually died) with his sister Caroline who followed him from Germany.

In Bath, Herschel earned a comfortable living writing music, organizing concerts and giving private music lessons. Herschel was interested in science and was an avid reader of scientific books taking a particular interest in astronomy. Having read about the 'charming discoveries' that had been made with the help of a telescope he decided that he wanted to see the heavens with his own eyes using a telescope. Unfortunately he did not have enough money to purchase a good telescope so he decided to build his own.

In typical Herschel fashion he read several books and then set about making the mirror for his reflecting telescope. Turning the entire house into a workshop he set about melting the copper and tin to form an alloy with which to cast the metal discs which would eventually be ground and polished into a mirror.

Failures would be tossed into the back garden and sometimes, the flagstones in the kitchen would explode when the molten mixture accidentally fell onto the kitchen floor. Once he had perfected the casting process, he set about grinding and polishing the metal discs into mirrors suitable for use in a telescope. He made over 200 mirrors and at one time spent 16 hours without stop, continually grinding and polishing the same mirror whilst being spoon fed by his sister Caroline so that he could keep going. Having made a mirror that satisfied his exacting standards (he tested them by inspecting the reflection of a candle flame to see if it was distorted) he set about building the rest of the telescope – including the eyepieces. Once the instrument was built he set about the task of his 'review of the Heavens' which consisted of an eight year project of familiarising himself with the night skies. Once he was familiar with the stars he realised that one 'star' appeared visibly bigger than the rest and suspected that it was a comet. When other astronomers heard of this find they too looked at it and decided that it was not a comet after all, but a new planet. Herschel named it after King George III (probably in an attempt to obtain a royal reward) but for some reason astronomers on the continent were not happy to name a new planet after an English king - so they named it 'Herschel'. Eventually it became known by the non-controversial name of the Greek god, Uranus.

Uranus was the first planet to be 'discovered' by a person. The other six (one has to include the Earth in the 'known' planet count) had been known about for all time because they could be seen by the naked eye. Herschel did receive his royal pension and gave up music to become a full time astronomer and telescope maker. His sister Caroline continued as his assistant and became the first woman to discover a comet receiving her own salary of fifty pounds per year from the King.

Herschel believed that all the bodies in the solar system were inhabited by intelligent life. He had turned his telescope onto the surface of Mars and he too had seen the darker areas as oceans and the lighter areas as land. He also believed that the Sun had a hard centre inhabited by little yellow Men. Around the hard centre were two clouds surrounding the entire core, or so he believed. The outer cloud was on fire and gave us the heat and light that we normally expect from the Sun; whist an inner cloud separated the inhabitants living on the core from the intense heat radiated from the hot 'outer' cloud. The middle opaque layer also prevented us from seeing the 'little yellow men' from down here on Earth. He did not see any problems in living on the Sun and thought that conditions would be similar to those found near the Earth's equator.

So it was that Herschel developed an interest in examining the Sun, and in 1800, Herschel set his mind to the problem of the spectrum produced when Sunlight passed through a prism. He was well aware that the Sun produced heat as well as light and he wanted to know if the colours of the spectrum carried this heat equally or if there was one colour only, which was responsible for all of the heat from the Sun. He arranged a prism so that it produced a decent sized spectrum (such that he could fit a thermometer into a single colour) and measured the steady temperature within each colour band. He used three thermometers in all, each one having a blackened bulb, as he knew that black absorbed all the radiation that fell on it and reflected none. Two of the thermometers were used as 'controls' – placed above and below the spectrum to measure the ambient room temperature where no light fell. He noticed that each of the colours gave a higher temperature reading than the control temperature readings but that as he moved further and further towards the red end of the spectrum the thermometer gave a higher and higher reading showing that more and more heat was carried by these colours. He then moved the thermometer into the dark region beyond the red end of the spectrum and found that this region was hotter still! Some reports say that Herschel only noticed this because he accidentally placed a control thermometer there! Herschell went on to show that these rays behaved in exactly the same way that light does being reflected, refracted and so on and decided that they were a part of the spectrum that our eyes are not sensitive to. He called them 'calorific' or heat rays but we now call them 'infra red' radiation, the 'infra' meaning 'beneath'.

Hearing of Herschel's discovery, in 1881 Johann Ritter, an ex-pharmacist and science/medicine graduate in what is now present day Poland, wondered if any kind of radiation existed at the opposite end of the spectrum, 'beyond the blue' so to speak. He was aware that certain salts of silver turned black when exposed to light. Boyle had investigated this effect much earlier and had come to the conclusion that the salts turned black because of oxidation due to their exposure to air. By Ritter's time, a great deal of work had been done to find the real cause, not because of its great potential in the future of the science of photography, but because it represented a way in which spies could communicate with each other as it was known by then that the blackening was caused by exposure to light. The silver salts, under the action of light, being converted into pure silver, which is black, causing the darkening. This is why, when you polish your silver trophies the cloth appears black, not because the trophy is very dirty but because some silver is transferred from the trophy to the cloth. It is silver that produces the 'black' in 'black and white' photographs.

It was also known that red light produced hardly any effect at all, the biggest effect being produced by blue light. Well, I think that you have gotten the idea by now, Ritter produced a spectrum from the Sun and placed some silver chloride salts in the dark area beyond the blue end of the spectrum and the silver chloride turned much blacker than it had done anywhere in the visible region. Hence 'Ultra Violet radiation' was discovered; Ultra Violet radiation having the meaning, 'radiation beyond the violet'.

For some time they thought that the three types of radiation were separate entities, with infra red and ultra violet just accompanying the visible light, but with the future discoveries of X Rays, Gamma Rays and Radio waves it was realised that they all formed a family of waves known as 'electromagnetic radiation'. All these waves have the properties that they consist of oscillating electric and magnetic fields, they can travel through a vacuum and when they do so, they all travel at the same speed – the speed of light (300,000,000 metres every second).

We gather information about the Heavens using every type of radiation. There is nothing special about light, the only thing that makes visible light 'special' to us is that it is the only part of the electromagnetic spectrum that our eyes have evolved to be sensitive to. In fact, what we know as light only makes up a tiny part of the whole spectrum. Our world would appear very much different if our eyes had evolved to be sensitive to Infra Red or Ultra Violet radiation. Can you imagine what it would look like?

Chapter 6. Bar codes and bunsen burners

You can learn a lot from something that doesn't happen - look at Sherlock Holmes. When the dog didn't bark in the night he knew who had committed the crime, not because of something that happened but from something that should have happened, but didn't!

On our road to the Big Bang, we will see that the 'things omitted' from the spectrum, tell us more about a star or galaxy than the 'things included' in that spectrum. Newton, Hooke, Herschel and Ritter had shown that white light is not a single entity but consists of a whole spectrum of colours, each colour being the way our senses respond to the stimulating effect of differing frequencies of light falling upon the retina of our eyes. The spectrum from distant stars is, or should be continuous (containing all the colours in the rainbow) but this is not the case. It has gaps or 'dark lines' interspersed within the colours. Each dark line within the spectrum represents a certain colour or frequency of light that is 'missing'. It is these 'missing' colours or missing frequencies of light that tell us a great deal about the star or galaxy that emitted the original spectra.

In 1802, William Hyde Wollaston, confirmed Ritter's discovery of Ultra Violet light. Wollaston, the son of a local clergyman, was born in Norfolk, England in 1766 and went on to study medicine at Cambridge. He did not find a career in medicine particularly stimulating and so he took up science, developing a technique to process precious metals. This brought him a substantial income thereby allowing him to study any other aspect of Science that took his interest. He was something of a 'Robert Hooke' character in that he didn't always follow things through to the end, dabbling with whatever aspect of science took his fancy at that particular time. But why not, when he had all that money coming in?

One invention that Wollaston is famous for and is in use to this day is a prism that splits a light beam into two separate rays – a beam splitter in other words. He used this device in his invention of the 'camera lucida'. This is a device that allows artists to cheat! When an artist wants to draw a scene, he takes the camera lucida with him and points it at the scene he wishes to paint. A piece of paper is placed on the base and, when he looks down on the paper, he sees an image of the scene projected onto it. All he has to do is draw around the outlines and there he has a perfect sketch. With all Wollaston's interest in prisms it is not surprising that he allowed sunlight to fall onto a prism to see the spectrum of light that it created. Wollaston succeeded were others hadn't because he firstly passed the light from the Sun through a narrow slit and then allowed the narrow chink of light to pass through the prism forming the spectrum. Wollaston saw the rainbow colours of the solar spectrum but what he also saw included amongst the colours, was a series of dark lines in the spectrum. Wollaston decided that these dark lines kept the colours of the spectrum segregated and separated the different colours i.e. he thought that there where certain colours in the rainbow and that these dark lines represented the boundaries between them. He was totally wrong in doing so as the spectrum is continuous and changes gradually from one colour to the next. There are an infinite number of colours and it is only ourselves who classify them as Red, Orange, Yellow and so on. What Wollaston had discovered, but did not realise, were 'absorption lines'.

Newton did not see the absorption lines in the Sun's spectra because, instead of passing the light through a slit, he passed the Sun's light through a hole in the shutters of his room. This gave several overlapping spectra from the Sun on the far wall instead of one single clear spectrum. The dark lines cancelled as the bright parts of one spectrum fell onto the dark parts of the other. Wollaston was happy to leave his discovery at that – after all he had found proof beyond doubt (or so he thought) of the existence of certain distinct colours and so he went on to spend his time and money on other topics that fascinated him. This left the field open to Fraunhofer to discover what the dark lines truly represented.

Josef von Fraunhofer was born in Bavaria in 1787 and was the youngest of eleven children. His parents died in quick succession and so by the time he was eleven years old, he found himself orphaned. His father had eked out a living as a glazier

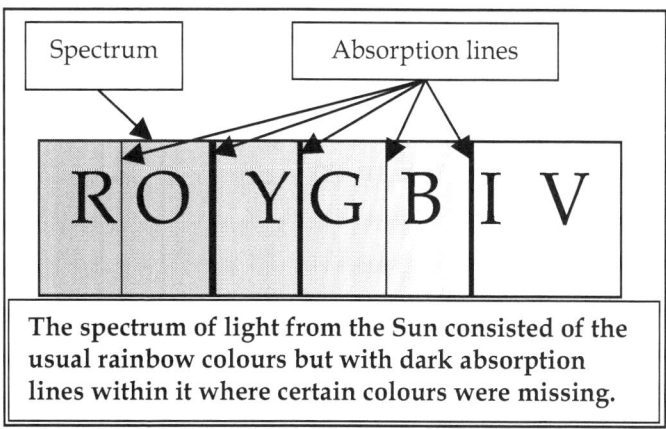

The spectrum of light from the Sun consisted of the usual rainbow colours but with dark absorption lines within it where certain colours were missing.

and since going to school was too expensive for his parents to afford, the young Fraunhofer had been forced to stay at home and help in the family business. Having learned something of the trade, it seemed sensible to let him continue. The young lad was bundled off to Munich to be apprenticed to his 'master', one Weichelberger, who worked with mirrors and glass. Unfortunately for Fraunhofer his master objected to any form of learning and Weichelberger refused to let Fraunhofer read books or attend any classes even on his 'off' days and so his chances of improving himself were nil.

You have probably heard of the expression that people who live in 'glass houses' should not throw stones - well this was certainly the case with Fraunhofer, as one day the whole 'glass house' fell down - completely! Due to structural problems, the building just suddenly collapsed upon itself and its occupants. His boss was dragged out fairly quickly but it took longer for the rescuers to locate Fraunhofer and drag him out of the rubble. As it turned out this was quite fortuitous for the boy and his luck suddenly took a turn for the better. During the time that he was lying under the pile of wood and bricks, not only had the usual 'ghouls' come along just to watch and stare, but also the future heir apparent, Prince Maximillian Joseph, came to see what all the commotion was about, and the prince, along with a local businessman, Joseph Von Utzschneider took over the rescue operation. Consequently, as the boy's saviours, they took him under their wings for the future. A very lucky 'non escape' one could argue as if he had been pulled out straight away by the crowd he would have missed his two benefactors altogether and all the benefits brought about by the paternal feelings in his two wealthy rescuers.

The Prince invited him for tea and gave him what was then a considerable sum of money for a young boy, which allowed Fraunhofer to buy himself out of his apprenticeship. Utzschneider gave him a collection of science books to read and later gave him a job at his optical works. Fraunhofer was good at his job and very soon took over the whole works, first as the manager and then as a partner and he turned the firm into the World's leading optical glass manufacturer, specialising in telescope lenses.

In order to standardize production, he needed to standardize colours. Not all people agree as to "which colour is which", as colour is perceived mainly from what your mother taught you. When you were young and learning to speak, your mother would point at a red toy train and say "r-e-d'. Eventually you would reply 'r-e-d' and Mummy would be very happy and give you a big hug as a reward. Now I am not suggesting that anyone has had a 'sad' Mum who, just for a laugh, would point at a 'r-e-d' train and say 'b-l-u-e-' but there are lots of shades of red and what might be cerise to you might be pink to someone else. This makes it difficult to have long and meaningful discussions on colour if the two of you are ever to agree.

Fraunhofer decided to standardize the colours of light scientifically by using the spectrum of the Sun. He knew of Wollaston's experiments and how Wollaston had described the dark lines separating the colours in the spectrum but what Fraunhofer did was to take a small telescope from the works and view the spectrum through this. What Fraunhofer saw were over five hundred lines in the spectrum from the Sun. He saw that, no matter if it was cloudy or clear. Morning, noon or evening, the lines stayed in exactly the same place in the solar spectrum, even though the actual brightness of the spectrum may change. In other words, their position in the spectrum was fixed. These dark lines or 'absorption lines' in the Sun's spectrum are named after Fraunhofer and are called 'Fraunhofer lines'.

If you are like me, then you are probably reading this book in the bathroom (I have a friend who has a special library of books in the bathroom solely for this purpose and it consists of books made up of 'short reads', with a few longer reads for special occasions). If this is the case, then pick up the toothpaste tube and look at the 'bar code' affixed to it. Imagine the bar code label coloured like a rainbow, red at one end going to blue at the other end. The black lines of the 'bar code' printed over the spectrum give you a good idea as to what the absorption lines in the solar spectrum look like; that is - they are a series of dark lines on a bright background.

He turned his 'spectroscope' onto the Moon and the planets and saw exactly the same pattern of dark lines in the spectra of the light given off by them. This suggested that the Moon and planets only reflected light from the Sun and did not emit their own light. When he turned his spectroscope onto the brightest star in the night sky, Sirius the 'Dog Star' he still saw the black lines. However, they were not in the same places as those seen in the spectrum of the Sun. Fraunhofer correctly reasoned that these dark lines had something to do with the star itself but left it at that and investigated no further. After all, he had succeeded in his original task of standardizing colours and regardless of what was actually causing the black lines, he could still use them to identify standard colours in the solar spectrum.

Fraunhofer went on to make his name in the World and the telescopes manufactured in the Munich factory went on, in the hands of others, to make great discoveries that would not have been made until much later, had it not been for Fraunhofer. When he died, he was given a state funeral and a medal was struck in his honour. For us though, it is time to move on our road to the Big Bang and bring astronomy back down to Earth.

In 1752, Thomas Melvill, a Scotsman, had carried out a series of experiments whereby he placed various substances into the flame of an alcohol burner. Different substances produced different coloured flames and to investigate this further he passed the light given off by the substance through a prism and examined the resulting spectrum. He saw that the spectrum was not complete but often had dark patches in it. If he stole some salt from the kitchen table and placed this in the flame, then it glowed a distinct yellow. When he examined the spectrum produced, by passing the light emitted by the burning salt through a prism, instead of seeing the familiar rainbow spectrum, all that he saw was a patch of yellow light with no red, blue or green.

Some time later, our old friend Herschel toyed with the same idea and proposed that this was an excellent way to detect small traces of impurities in certain known substances. Fraunhofer and others had noticed that certain strong dark lines in the Sun's spectrum coincided with the yellow lines given out when sodium was placed in a flame and concluded that sodium must exist in the atmosphere of the Sun. However, it was Robert Bunsen and Gustav Kirchhoff who put spectroscopy, as it became to be known, on the map.

Robert Bunsen, the son of a University linguistics lecturer, was born in Germany in 1811. He had always had a burning ambition to study science and Chemistry in particular. He had made a name for himself studying cacodyl and its derivatives. Cacodyl is Greek for 'the most obnoxious smell you have ever smelled in your life' and that is what it does, stinks the place out. You only need one drop of it in a room and that is it; you have to leave. Bunsen himself complained how the stench affected him and gave his tongue a black coating. Hardly surprising, as the salts not only contained arsenic but they were also highly explosive. On one occasion in the laboratory, the mixture in question exploded sending a shard of glass into one of Bunsen's eyes with the result that he eventually lost the eye. Another of Bunsen's noted scientific discoveries was in geology and the explanation of how geysers work. After making tests on a volcanic eruption that occurred in Iceland, he concluded that the water boiled on its way to the surface and not deep underground in the Earth. As the hot water gushed up the gaps in the rocks, the pressure fell and since the temperature at which water boils depends upon the pressure, the hot water would reach a point where it would boil, change to steam and squirt the column of water above it high into the air. This is the same effect as that which happens when your car overheats – do not open the radiator cap whilst the car is hot to check the water level because, as you open the cap to release the pressure, the water boils and you will discover your own personal geyser! To check his theory, Bunsen built a model geyser in the laboratory and found that it worked to perfection including the rumbling sounds that fanfare the oncoming spurt of the geyser itself.

Gustav Kirchhoff was born in 1824 in what is modern day Russia. He was the son of a lawyer and showed a great intellect as a child. Kirchhoff is renowned for his two laws on electrical circuits, namely:

- All the current that enters a junction leaves it.
- The voltages across all of the components in a circuit add up to the voltage across the battery.

These laws are learned by just about every school child studying Physics at school but he derived these laws whilst he was still an undergraduate. On graduating, he married the daughter of his mathematics professor and eventually ended up working with Bunsen in the university at Heidelberg. They had met previously and become very good friends and it was Bunsen who had persuaded Kirchhoff to move to Heidelberg so that they could work together. It was not all plain sailing though and one of Kirchhoff's blunders, along with that of another scientist Weber was that they had both found, independently, that the speed at which an electrical current travels along a wire had nothing to do with the wire itself and was equal to the speed of light. They both put this down to a coincidence of the numbers thus leaving the gates open for Maxwell to make the link between light, electricity and magnetism some five years later.

So it was that Bunsen and Kirchhoff had teamed up to investigate the spectra emitted by different elements when heated in a flame. However, the heat sources that they had at the time were insufficient for detailed scientific work. They were not bright enough, the flame flickered and the flame also had a strong colour in itself, bright yellow from the sodium. Bunsen set about finding a suitable light source. He took a design put forward earlier by Michael Faraday and perfected it, getting his laboratory technician to put the finishing touches to the final burner. The burner, which is now named after him gave a bright colourless flame and this enabled one to see the colours of light emitted by the elements placed in the flame. Kirchhoff managed to produce far purer substances than those before him and he also made the spectrometer – an instrument with which to inspect the emitted spectra.

John Draper (we met him before if you remember - it was his legacy that enabled the Harvard Observatory to complete the 'Draper Catalogue' of stars) had already determined that a hot, solid body would give a continuous spectrum whilst a hot gas gives a discontinuous spectrum containing dark bands and spaces. Bunsen and Kirchoff confirmed Drapers results and put spectroscopy onto a firm footing by showing that each element had its own characteristic spectrum that was unique to that element. That is, if you looked at the emission spectra (which colours of light are given out by a particular element) then one could tell which element it was. They also showed that when white light is passed through a gas, then the colours of light that are absorbed by the gas are the very same colours that are emitted by that gas when it is heated. Using these techniques, Bunsen and Kirchoff had detected earthly elements in the outer layers of the Sun and they suggested that by looking at the spectrum of the Sun, one should be able to determine its entire composition.

It is ironic that Bunsen, who made so many discoveries in science, is known today for a burner that he only perfected from an original idea of Faraday and one that his laboratory technician built (and no doubt came up with the idea of the familiar collar and two holes in the base of the tube). That was the great thing about the burner, the gas was mixed with the air before combustion took place and not during, as with other burners. The burner was so successful that orders came in from all over the World and so the technician's son set up a company to manufacture them. Since they never patented the burner, it was soon copied and sold by others – some of whom tried to patent it themselves, but were unsuccessful in their attempts.

Chapter 7. Coffee in a lay-by.

On our journey to the Big Bang, this is a good point to pull over into a lay-by to buy a cup of coffee from the trailer set up as a coffee shop. We need some background information so that we can understand what is being discovered and why it is happening. If you already know all this, then feel free to skip this section and move onto the next chapter, but I will try to make it interesting as well as informative. What we need to know is how a continuous spectrum (like the rainbow) and line spectra such as absorption and emission spectra, are formed.

We have already seen how Hooke, Huygens and Newton almost came to blows over whether the nature of light is a particle or a wave. Hooke and Huygens rooted for a wave whilst Newton fought for a particle theory. Generally, Newton won the day - partly because of his outstanding intelligence and standing in the scientific world and partly because Hooke 'popped his clogs' so to speak and died leaving the field open to Newton.

Newton's model of light was one of particles or corpuscles – billions of perfectly elastic little particles given out by light sources. These travelled out in all directions, bouncing off objects so that the angle of incidence was always equal to the angle of reflection, changing speed as they enter glass to account for refraction, each colour having its own type of particle to explain the different colours of the spectrum and so on until he had just about explained every phenomena known at that time. Newton's argument against the wave theory was that no one had seen the diffraction of light waves and diffraction is something that all waves do.

Diffraction is the spreading out of a wave as it passes through a gap or past an object. Consider sitting in a room with the door open. You can hear people out in the corridor because as the sound waves pass through the open doorway, the sound waves diffract or 'spread out' so that they completely fill the room. One can hear the people outside in the corridor, no matter where you are sitting in that room. Light, on the other hand, does not diffract as it passes through the doorway; it travels in straight lines. In order to see anyone outside in the corridor, they have to be in a direct line of sight otherwise we do not see them. We know that this is because the wavelength of light is far too small to obtain appreciable diffraction in everyday life. In order to get a decent spreading of a wave, the wavelength must be about the same size as the width of the gap (the doorway in our case here). With sound waves or water waves, this is not a problem because their wavelength is usually a few metres and there are lots of gaps of this width, including the doorway. The wavelength of light is around 0.0005 of a millimetre and in order to achieve an appreciable amount of diffraction you are looking for the light to pass through a slit or gap having a width of approximately this size. This is about the width of the scratch that would be made by a brand new razor blade if it were drawn along a painted surface. Needless to say, that if you did find an opening of this size, then so little light would pass through it that you would not see anything - as it would be too faint to be noticed. Remember that in Newton's days the only sources of light were candles and the like – they did not have lasers as we have today. Since light was not seen to diffract, Newton insisted that it was not a wave. Had he known about the diffraction of light then things may have been different. But wait a minute, Newton did know about interference; what about Newton's rings? This is where Newton had placed one lens on top of another and he had seen coloured rings caused by the various rays of light overlapping and interfering. The answer to this question is that Newton ignored this result because he felt that he did not need it in order to support his theory on light. That is, he felt that he had enough evidence with his spectrum formed by the prism to prove his theory correct and did not need any extra evidence. Or did he just ignore the effect because his theory could not explain it?

One observation that Newton never managed to explain convincingly was that two beams of light could be shone across one another and the beams continue as if they had never crossed. If light was a particle, then surely the particles must collide and bounce off each other to produce a dispersion of the beams. The only things that can cross each other without having any effect are waves. However, Newton 'fudged' his way past this one and his corpuscular theory of light ruled as 'king' during his lifetime.

Newton died in 1727 and seventy-four years later, in 1801, Thomas Young demonstrated the interference of light with his 'double slit' experiment.

Young was born in England in 1773 and as a child was somewhat of a prodigy, studying as many as ten different languages. At University he studied medicine and just to prove that 'money goes to money' or 'the more you have the more you get', he was left a house in London and a huge fortune by his Uncle. This allowed him to do as he pleased – and he pleased to study Science. He did set up a medical practice in the London house but followed his other interests at the same time. His interests were many and varied and included the explanation of how the human eye produces an image, the stretching of wires and his 'Young's Modulus', the relationship 'work done is equal to force multiplied by distance', and the term 'energy' itself. He also played a part in the deciphering of the 'Rosetta stone'. The Rosetta stone was a stone found in Rosetta, Egypt by Napoleon's army in 1799 and proved to be the key to understanding ancient Egyptian hieroglyphics – those funny symbols that adorn Egyptian monuments and buildings. A French soldier saw the stone and decided that he wanted it as a 'souvenir' and took it back to France where it became a family heirloom – no mean feat, as the stone is about four foot by two foot and is heavy! The stone was eventually sold in a flea market in Paris where an Oxford professor who studied Egyptian artefacts just happened to be passing on his holidays. He bought the stone and took it back to England where none of his colleagues could understand a word of it and so it was left to gather dust in a cellar. Many years later, a nosy undergraduate student took another look at the stone and realised that it contained the same message written in two different languages but three different scripts, Greek, Demotic and Hieroglyphics. The Greek was easy to translate but not the rest. It was then that Thomas Young had a look at the stone and, with his ten languages learnt as a child, he realised that there was a pattern between the Demotic and Hieroglyphic nomenclature. The Frenchman Jean Champollion followed on from Young's work and succeeded in showing that the Hieroglyphics formed a spoken language. From this day on, all Hieroglyphic inscriptions could be deciphered. What did the inscription read? Well it was fairly boring in itself, just a load of instructions to tourists as to how they could pay their respects to a certain king, a sort of 'take your shoes off before entering' message in three different languages.

Coming back to Young's investigations on light, he demonstrated the diffraction of light by passing it through a narrow slit and noted that bright and dark fringes emerged from the other side. He also passed light through two narrow slits, side by side, and showed that interference occurred. Interference is where two waves arrive at the same place at the same time and 'interfere' with one another. If the crest of one wave meets the crest of another wave, then a single wave results but it is now twice as big as the two single waves that formed it. If the crest of one wave meets the trough of the second wave, then the two waves cancel and there is no resulting wave. The interesting point here is that one light wave can be shone onto a surface and it is illuminated. Shine a second light wave onto the same surface and they can cancel, if the conditions are right, and instead of giving a brighter area, give darkness! Interference is a wave phenomenon – if interference occurs then it is waves, not particles and so Young had shown Newton wrong and Hooke correct in that light was a wave and not a particle. Pity that Hooke wasn't still around to gloat! However, it still took many years before the wave theory of light became generally accepted. The 'Coup de resistance' for the wave theory came in 1850 when Foucault measured the speed of light in glass - and found it to be slower than in air. Just as the back end of a high powered car swings around when you put your foot down too hard on the accelerator and speed up too quickly, Newton's corpuscular theory had the corpuscles speeding up as they entered the glass and thus changing direction. When Foucault showed that in reality exactly the opposite was happening, the corpuscular theory had to go.

Never the less, it did not take long for the pendulum to swing back. The particle theory of light proposed by the Greeks and

Newton came back to the forefront with two experiments that heralded the birth of quantum mechanics.

In 1887, in Germany, Heinrich Hertz showed that when he transmitted radio waves across his laboratory, sparks could more readily be obtained across a small gap between two wires, if he shone ultra violet radiation onto the wires. One year later, a second German Physicist, Wilhelm Hallwachs, investigated this phenomenon further and found that a negatively charged, clean zinc plate rapidly lost its charge when ultra violet radiation was shone onto it.

This led to the understanding that radiation is not a continuous quantity, where it is possible to have any amount, but that it is 'quantised'. Being 'quantised' means that the energy comes in 'chunks' of a certain size and you either have one, two or three chunks of it, but you cannot have anything in between like 'one and a half chunks' of it. There is nothing surprising about this as it happens in nature all of the time, just like the number of flowers on a plant. There are either one, two or three flowers on the plant and never 'one and a half' or two point seven flowers'; you see, nature is quantised. The discovery and development of Quantum Mechanics can, and has taken up several books and so to follow it here would be too big a diversion from our road to the Big Bang. We will only look at the signpost showing the way and see where the diversion leads.

When we go to the seaside and look at the sea, we can see the waves coming towards the shore. The water waves are continuous and stretch as far as the eye can see. Hooke imagined light waves to behave in just the same way but we find that this is not the case. With light waves, the energy is not spread out and evenly distributed but it is concentrated into chunks or packets or quanta of energy called 'photons'.

The energy of each photon depends on the frequency or colour of the light and they are related by the formula:

(Energy of photon, E) = (Planck constant, h) x (frequency, f)
Or simply;

$$E = hf$$

The higher the frequency, the more energy each photon has, so photons of blue light have more energy than photons of red light. We are not talking about waves any more, we are saying that light is a particle (good old Newton). There are billions of photons given out by a light source every second and that is why we see an overall, even illumination. It is very much like spraying a wall with graffiti using a can of spray paint. The aerosol sprays out tiny blobs of paint which arrive in a dot, dot, dot fashion, but if one sprays for long enough, you will get an even, overall coating of paint. This is the way light works. When a lamp illuminates a surface, it is not all illuminated at the same time. Bits here and bits there of the surface are lit up after photon after photon arrives on the surface. There are so many of these photons that over a short period of time we see an overall, even illumination.

Talking about water waves, have you ever wondered about water waves? I mean really wondered about them? I was lying on a beach in Greece and suddenly realised that wherever I had been on any beach around the Mediterranean Sea, the waves travelled inshore. Be it North Africa, Spain, France, Egypt, the waves always travel ashore. That is, in any ocean or sea, the waves travel outwards – appearing to spread out from the centre. Furthermore, when I was lying on the beach in Cyprus, an island of which I am particularly fond, waves always came inshore there too, even though Cyprus is in the middle of the Mediterranean where all the waves are supposedly spreading out from - this, in order to wash upon all the surrounding beaches around the Mediterranean. This phenomenon is not restricted to the Mediterranean but happens in all the major oceans in just the same way. That is, wherever one goes the waves always come ashore.

The question is, "what happens in the middle of the sea?" If, in all the seas and oceans of the World, the waves travel inshore, (that is the waves spread out as if they all come from the middle of the body of the water) what happens in the middle? Does the water 'part' in the middle? If this is not bad enough, we are often told that these waves are caused by the wind blowing over the seas, so if the winds are blowing East to West, does that mean that waves should be driven onshore in the West but away from the shore in the East? Additionally, in the daytime the winds are 'sea breezes' and come inland. No problem with that - but at night, the winds change direction and become 'land breezes' where the wind blows out to sea. So, why don't the waves change direction and travel in the opposite direction, I mean travel away from the shore at night times? Have you ever seen waves travelling away from any seashore and if not why not? It was at that point that I gave up and went for a drink. Maybe the answer to this could be the basis of the sequel to this book!

Apologies for those ravings, let's get back to the quantum mechanics. Light consists of photons and the energy of each photon depends upon its frequency or colour. This means that some photons can do things that other photons can't – because, individually, they just don't have the energy. This is why, in the old days, photographers could develop their black and white photographs whilst working in a red light. Red light has a low frequency and therefore photons of red light do not have enough energy to expose the photographic paper. Other colours have a higher frequency and therefore these colours have photons of higher energy and thus each photon has enough energy to expose the photographic paper.

Further developments in quantum physics by Neils Bohr in particular, led to the idea that the atom itself was quantised. In Bohr's model of the atom there is a small central, positive nucleus with the electrons orbiting the nucleus in the same way that the planets orbit the Sun in our solar system, but it is not as simple as that. An electron going around in a circle is accelerating and we know that an accelerating electron radiates energy. That is, it gives off electromagnetic radiation (called bremsstrahlung – we will meet this radiation later). However, this cannot be so, since if the electron radiates energy then it will spiral into the nucleus in a very short time indeed. Bohr proposed that there were only certain stable orbits that the electron could occupy without radiating energy and that these stable orbits corresponded to a certain amount of energy that the electron could have, called energy levels. The less energy the electron has, then the closer it orbits the nucleus. It is actually far more complicated than this but this model will suffice for our purposes.

Usually, an atom has its lower energy levels filled with electrons whilst the higher energy states or orbits are empty. This applies to all life; things like to be in a state of lowest energy – this is why we fall over when drunk! When we fall over, we are undergoing a transition from a higher state of gravitational potential energy to a lower state of gravitational potential energy.

An atom can become 'excited'. This is where one of its electrons has somehow managed to gain just enough energy to jump up to a higher energy level. It could be due to a second 'free' electron coming along and bashing into the electron trapped inside the atom. In fact this is the usual way of doing it. We place a gas at low pressure in a narrow glass tube and put a voltage of a few thousand volts across it. There are always a lot of 'free electrons' about (we create thousands every time we light a match) and these free electrons are repelled by the negative terminal and attracted to the positive terminal. The free electrons accelerate and on their way they will collide with the bound electrons in the atoms of the gas. On collision, the bound

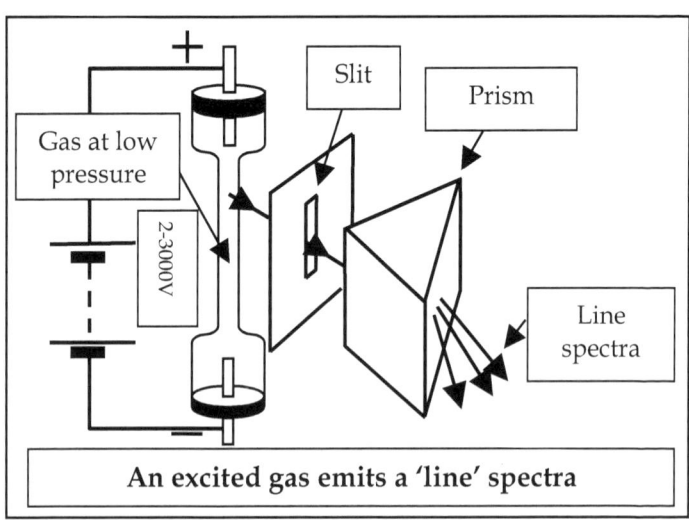

An excited gas emits a 'line' spectra

electron in the atom can gain just enough energy from the free electron to leap up to a higher energy level – but it can only accept exactly the correct amount, no more, no less.

Now the atom is not too happy in this excited state. It has a lower energy level that is missing an electron and it has an electron that has risen above itself and is occupying a higher energy level. What happens is that the electron falls back down to its original level, giving out its excess energy as a photon of electromagnetic radiation. The fluorescent lights in our home, work on this principle. The frequency of the radiation emitted is dependent on the difference in energy between the two energy levels of the atom. Consider an electron falling down from an energy level of energy E_2 to a lower energy level of energy E_1, then the frequency of the emitted radiation is determined by the relationship:

$$hf = E_2 - E_1$$

The bigger the difference in the energy levels, the higher the frequency of the emitted photon. There are many different energy levels that can be occupied by the electrons in an atom and an even larger range of jumps that can be made by the electrons from one energy level to another. This means that we do not get just one frequency of radiation emitted but a series of distinct and discrete frequencies of light given out. This produces our emission spectra given out by an excited gas. It is known as a 'line spectrum' because when the radiation is passed first through a narrow slit and then through a prism, the resulting spectrum is a series of coloured lines on a dark background. Since the frequencies emitted are dependent upon the energy levels within the atom, the arrangement of these lines within the emission spectrum is characteristic of that atom. That is, by examining which lines are present in the emission spectrum given off by an ignited gas, one can tell exactly which atoms of which element are present in that gas. Emission spectra are the gas's 'fingerprint' and no two elements have the same 'fingerprint' or emission spectrum.

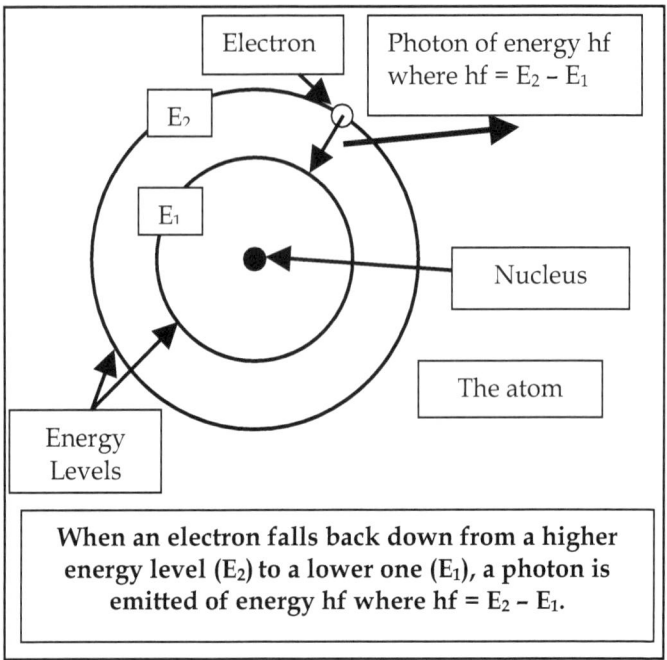

When an electron falls back down from a higher energy level (E_2) to a lower one (E_1), a photon is emitted of energy hf where hf = $E_2 - E_1$.

This only works for gases where the atoms in the gas are isolated and are so far apart that they do not affect one other. Their energy levels have one value only, take it or leave it.

In solids this is not the case; the atoms are packed closely together so that one atom affects the energy levels of the atom next to it. The result of this 'closeness' is that the energy levels broaden and can even overlap. Instead of the allowed orbit having a single radius representing a single energy level, we now have broad bands representing a broad range of allowed energies for the electrons. This in turn means that there are whole ranges of energies an electron can have, in order to move from one band to another - which results in ranges of frequencies in the emitted radiation. If a solid is heated and the emitted radiation passed first through a narrow slit and then through a prism, the spectrum will consist of patches of colour or even the entire rainbow.

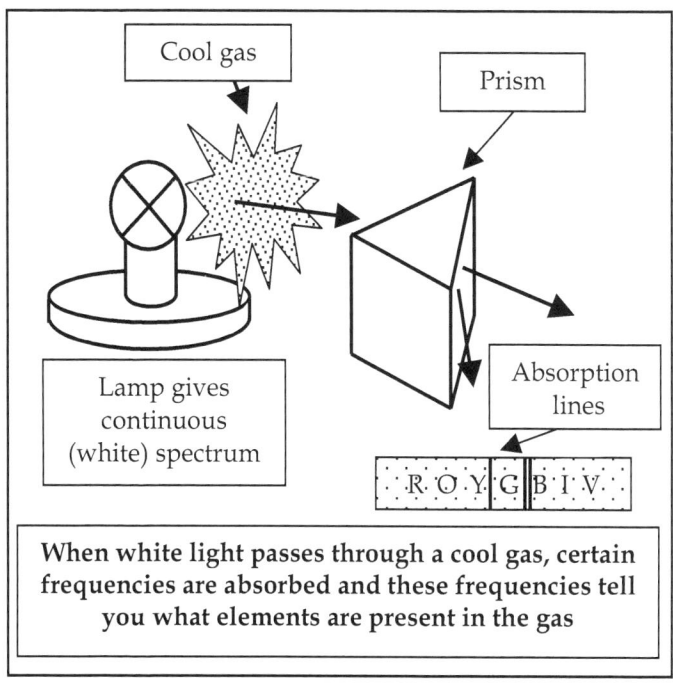

When white light passes through a cool gas, certain frequencies are absorbed and these frequencies tell you what elements are present in the gas

This is the difference between a 'line spectra' from heated gases and a 'continuous spectrum' from heated solids. If we take a continuous spectrum (from a heated solid) and pass it through a gas, the light that emerges will have certain colours of light missing and the colours that are missing correspond exactly to the colours that would have been emitted had that same gas been heated. The emerging spectrum consists of the usual rainbow but with 'dark lines' placed where those colours used to be. That is, a cool gas absorbs the same colours or frequencies of light that the same gas would emit if it were hot. This 'rainbow' spectrum with the characteristic dark lines is called the absorption spectrum of the gas.

These certain colours have disappeared because they have been absorbed in 'exciting' the atoms of the gas. The electron in the atom absorbs photons of light, having just the right frequency and energy to raise the electron from a low energy level to a higher energy level. Other photons just don't have the right amount of energy, corresponding to the difference in energies between energy levels within the atom, so when they bump into an electron they are re-emitted as a new photon identical to the first so that to all intents and purposes, the original photon has been transmitted.

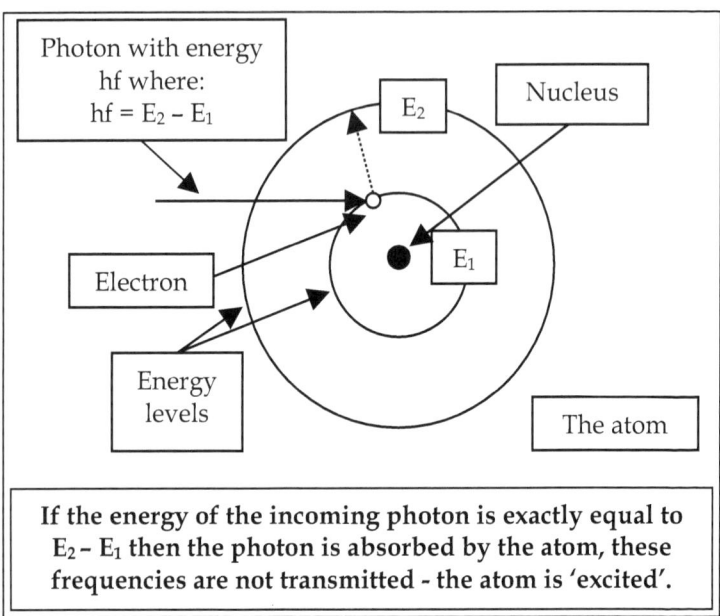

If the energy of the incoming photon is exactly equal to $E_2 - E_1$ then the photon is absorbed by the atom, these frequencies are not transmitted - the atom is 'excited'.

The emission and absorption spectra are characteristic of the element to which the gas belongs and Kirchhoff used this result to determine which gases are present in the atmosphere of the Sun. The analogy between absorption lines and bar codes on products one can buy in a supermarket is a good one. When the laser scanner at the supermarket checkout scans the bar code one can tell not only that it is toothpaste, but also which brand, price and size it is. Everything the cashier needs to know about that product is contained in that series of black lines printed on the white label.

In the hot dense interior of a star, a continuous spectrum is produced consisting of all the colours in the rainbow.

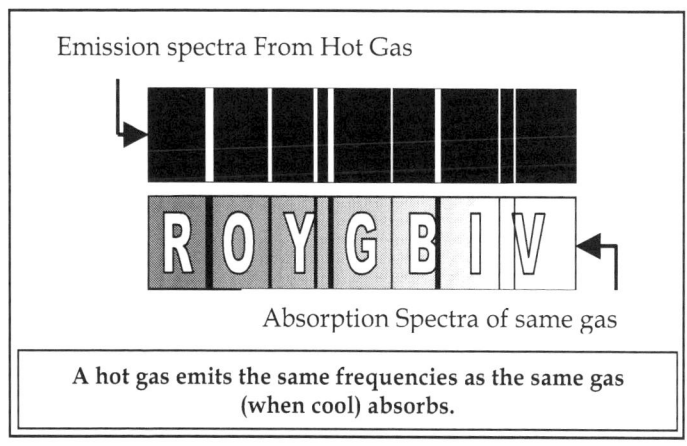

Surrounding the star is a mass of cooler gas and as the continuous spectrum of 'white light' produced in the interior of the star passes through this cooler gas, certain frequencies or colours of light are absorbed. The colours absorbed are characteristic of the gas surrounding the star and so the spectrum received here on Earth consists of the spectrum with a series of dark lines in it.

In the same way as the scanner at the supermarket, astrophysicists look at these dark lines and compare them with the emission spectra of known elements here on Earth. By matching the dark lines in the absorption spectra of the star with known emission lines of the elements, we can tell which elements are present in the gas around a star and assume that the star has these same elements inside it.

This is all the background information we need to know so, if we have finished our coffee, we will put the car into gear, look in the rear view mirror and pull out of the lay-by, and back onto the road to find the Big Bang.

Chapter 8. Trumpets and double stars.

The techniques for determining the composition of stars were now well developed and, as with all new experimental techniques, it was welcomed with open arms. The colour and spectrum of the stars now featured at the forefront of scientific research since this was a new area in which discoveries and reputations could be made.

It was known that some double stars had different colours and an Austrian Mathematician from Saltsburg set about solving the mystery.

Christian Doppler was the son of a stonemason but he was unable to carve out a name for himself in this area because, as a child he was forever sickly (have you noticed that thus far just about every scientist has either been sickly, orphaned or had something wrong with them as a child? Perhaps that is where I went wrong, I was perfectly normal and healthy!). Consequently, Doppler was sent to study Mathematics and became a teaching assistant at the local University whilst he tried to find a 'proper' job.

In order to secure a proper teaching job, the applicants had to sit a common examination and then teach a trial lesson, and Doppler tried and tried again to do this but with little success. Becoming thoroughly cheesed off with the whole idea, he decided to sell up and immigrate to the USA when, just as he was about to sell his last tea spoon, the offer of a job came through in Prague. He had applied and had been 'interviewed' for this post two years earlier, so they must have been very impressed with his interview technique (perhaps they where just trying to save money by 'interviewing' so many people that they had enough 'trial lessons' to keep their students happy without employing a paid teacher). Though it could be said, that the students may have been better off without him, because some of Doppler's students led an uprising - complaining that Doppler was failing too many of them in the examinations. The authorities looked into the complaint and took the side of the students with the result that they were all allowed to re-sit the examinations and Doppler was given an official reprimand – later withdrawn after he appealed.

During all this, in 1842, Doppler set his mind on the problem of 'how could double stars have different colours, shouldn't they be the same'? Doppler was helped by the fact that he refused to believe that light was a transverse wave. He was convinced that light needed a medium to travel through and that light was a longitudinal wave behaving in the same way that sound waves behaved (like 'shock waves' where the particles of the medium vibrate along the direction in which the waves travel and not like water waves, where the particles vibrate up and down whilst the wave travels along). This was the time of railways and Doppler was aware that the note given off by a train as it sped past people on the trackside changed in pitch as the train passed. The noise of the train approaching has a high pitch, whilst once the train has passed, the noise falls to a low pitch. Doppler argued that it would be the same with the stars. Light emitted from a pair of double stars rotating about a common axis will differ as the light from a star moving towards the Earth would appear to have a higher frequency and thus appear 'bluer' whilst light emitted from a star moving away from the Earth would appear to have a lower frequency and thus appear 'redder'.

The velocity of a wave in a medium - for instance sound waves in the air or water waves on water, is independent of the velocity of the source or the observer. The velocity depends upon the conditions of the medium itself. It does not matter if one is standing still or shouting from inside an open top car, the sound waves, once emitted will both travel at the same speed and it is this principle that is responsible for the shift in wavelength of the wave.

Imagine a duck, at rest, floating on the surface of a duck pond. If the duck decides to prune its feathers, circular ripples will be sent out on the water surface equally in all directions. No matter where we are around the edge of the pond, the velocity, frequency and wavelength of the ripples will be the same.

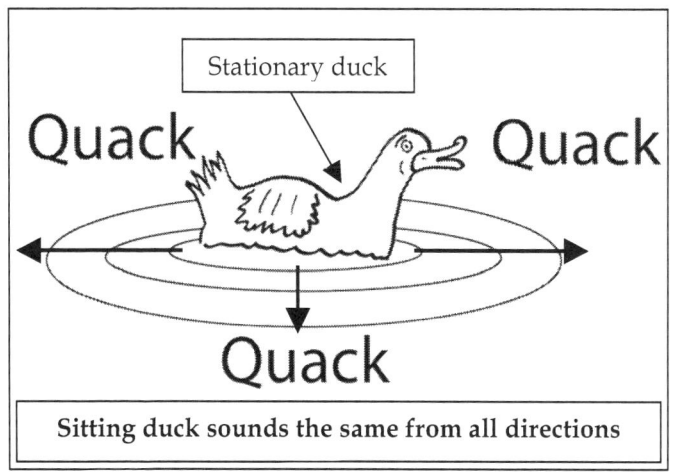

But what will happen if the duck is swimming along? The velocity of the waves will be the same as before but the waves will be squashed together in the region in front of the duck and stretched apart in the region behind it. A person standing on the banks of the duck pond will now measure different values for the wavelength and frequency, depending on where they are stood.

A person standing in front of the duck will receive waves of a shorter wavelength and hence a higher frequency (since the wave velocity is equal to the frequency of a wave multiplied by its wavelength and the wave velocity is a constant in the pond). We say that the waves have been 'blueshifted' because, in light, a shorter wavelength means a higher frequency and that means that the wave has been 'shifted' towards the blue end of the spectrum. If a star is moving towards us, the light waves received here on Earth are squashed and will have a shorter wavelength, shifted towards the blue end of the spectrum.

A person standing on the bank of the pond behind the duck will receive waves of a longer wavelength and lower frequency, since the movement of the duck is such as to stretch these waves. We say that these waves have been 'redshifted' as in light, waves from a star moving away from us are stretched and have a wavelength shifted towards the red end of the spectrum.

The faster the duck is moving, the more the waves are stretched or squashed. By measuring the change or 'shift' in wavelength, one can calculate how fast the duck is travelling.

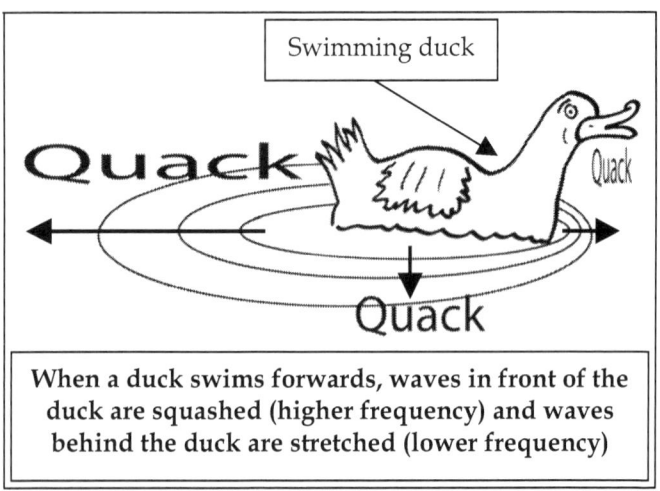

When a duck swims forwards, waves in front of the duck are squashed (higher frequency) and waves behind the duck are stretched (lower frequency)

Doppler did his sums (which was not easy for him as it is alleged that he wrote a basic mathematics text book that was full of errors) and he worked out a formula by which the velocity of the duck or moving object could be measured which is; the velocity of the moving object, v divided by the velocity of the wave in the medium, c is equal to the shift in wavelength, $\Delta\lambda$ divided by the original wavelength, λ:

$$\Delta\lambda/\lambda. = v/c$$

Or,

$$v = c\Delta\lambda/\lambda$$

This is the principle of many roadside speed traps. These speed traps send out a wave of known frequency and it bounces off the front of your car. In reflecting this wave, your moving car squashes the waves together, the faster the car is going the more the waves are squashed together. The speed trap receives the waves, measures how much they have been squashed, works out how fast you were going and sends you a speeding ticket.

However, getting back to the duck, a person standing at the side of the pond will see the duck moving across his line of sight. He would not notice any difference in the wavelength of the waves arriving from the duck. In fact, from the waves alone, he could not tell if the duck was moving or not. This is a fundamental point in the Doppler Effect; you can only measure the 'radial velocity' of the moving object; that is, one can only measure the velocity of the object directly towards or away from you. One cannot measure the actual velocity of the object itself.

This is one way of applying astrophysics to everyday life. When one is speeding along the highway and notices a speed trap in the distance, astrophysics suggests that instead of keeping going in a straight line or braking quickly, it might be a good idea to swerve. That is, swerve so your path is now a perfect circle with the speed trap at the centre. Your 'radial velocity' is now zero towards the speed trap and that is what it would record, zero. No matter how fast one was going! Please don't try this, I am only joking; drive at a safe speed, always.

The Doppler Effect, as it is known, applies to distant stars and galaxies and Doppler predicted that it actually would be used in the future to measure the radial velocities of distant stars. In order to prove his theory, Doppler thought of a novel experiment. He could not test his theory on light, as terrestrial shifts in frequency and wavelength would be too small, so he decided to test the theory on sound. Since there were no frequency meters to measure the frequency of the sound in those days, Doppler used two groups of trained musicians, or trumpeters to be exact and used their trained musical ears to measure the pitch or frequency of the sounds.

He put one group of trumpeters on an open railway truck and asked them to play a certain note constantly. Next, he arranged for a train to pull the trumpeting trumpeters past a second, stationary group of trumpeters at the side of the track and it was their job to determine the pitch of the note both as the train approached and as it disappeared. Like all good scientists, Doppler repeated his results and so they spent the day with the train going up and down the track at various speeds, with trumpeters a trumpeting, whilst those on the side of the track determined the apparent pitch of the note as they heard it, when stationary on the ground.

However, with this simple experiment, he was able to prove his results. But, what happened to his theory on the colours of the double stars? Completely wrong! The effect on light is far too small to completely change the colour of the stars; the stars are just different colours.

Mention the name of William Huggins to a group of astronomy students and they will probably think that you are talking about the troll in Tolkein's "The Hobbitt" but it is also the name of a famous scientist. Huggins was born in England in 1824 and he took an interest in astronomy early on. He had been to Cambridge but he dropped out in order to look after the family drapery business. Not being happy in this role he sold the lot and set up his own observatory just outside London. He married Margaret Murray, an Irish lady who followed the same interest in astronomy as Huggins and together they investigated the Heavens.

Huggins had heard of Kirchhoff and Bunsen's work in identifying terrestrial elements in the Sun and of their suggestion that one should be able to determine the composition of the Sun by examining its spectrum. He set himself the task of doing just that, both on the Sun and on other stars. This was not an easy task, as the light received from other stars is miniscule when compared to that received from the Sun. One must also remember that at this time, photography was not sufficiently developed as to be of any use. Emission spectra of known gases had to be produced in the observatory itself so that a direct comparison could be made with the absorption lines in the spectra of the stars. Observatories started to take on the look of a science laboratory, as the huge batteries, discharge tubes and induction coils were brought to produce the emission spectra needed for comparing the two spectra.

To put it simply, Astrophysics was born.

Being a new subject, Huggins was virtually able to make a new discovery every night but his greatest contribution, as far as our road to the Big Bang is concerned, was in being the first to apply the Doppler Effect to stellar motion. It had taken over twenty five years for this to happen after Doppler's original prediction because until Kirchhoff, in 1859, had shown that terrestrial elements existed in stars, no one had the means to know what the original wavelength of the waves was – and this is a necessity in Doppler's equation. Huggins was able to do this. He pointed his telescope at a star, passed the light received through a prism and produced a spectrum. The spectrum contained the dark absorption lines characteristic of the elements around the star. He compared these absorption lines with the emission lines of known gases produced in his own observatory and this told him two things. Firstly, it told him what the original wavelength was, since he was producing that wavelength in his laboratory, and secondly, it told him by how much that wavelength had been shifted. Knowing the velocity of light, the radial velocity of the stellar object could be found.

Remember our bar codes on the toothpaste tube? Well this is the same thing once again. The bar codes are unique and give the cashier at the teller all the information that they need to know about that product. Now imagine if the machine printing the original bar codes had developed a fault. Instead of printing the bar codes nicely in the middle, it starts printing them shifted towards one end of the label. We do not have to throw these labels away as we can still use them. The bars are in exactly the same configuration as before so when the laser at that check out scans them, it will still know exactly what the product is. The bars are just shifted to one side of the label; we still know what it is.

It is the same with the absorption spectra produced in the light of a moving stellar object. The dark lines are still in the same configuration, so we know which element is responsible for creating them, but they are shifted to one side of the spectrum. If the stellar object is moving towards us, then the dark lines will appear nearer the blue end of the spectrum – blueshifted. If the

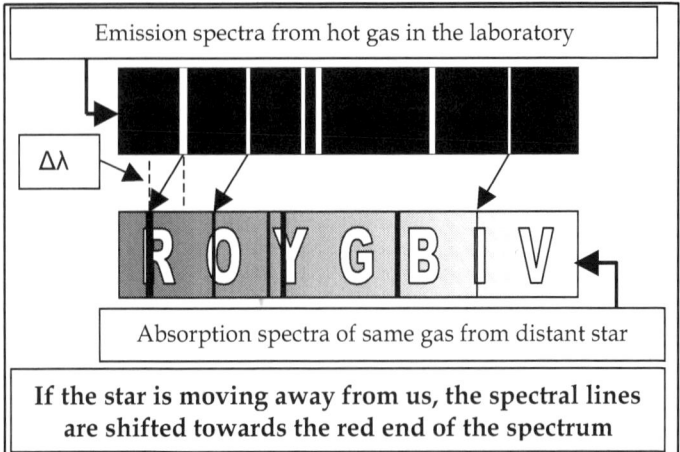

If the star is moving away from us, the spectral lines are shifted towards the red end of the spectrum

stellar object is moving away from us then the dark lines will appear towards the red end of the spectrum – redshifted.

Huggins, along with his wife, made a catalogue of the spectral lines of many substances and used these to determine the composition of many stars and nebula. He was the first to detect a Doppler shift in light from some distant stars and, in 1868, determined that the nearby bright star Sirius, was moving away from us at a speed of over twenty miles per second.

Whilst the science of astrophysical spectroscopy was becoming more widespread, the search for life on Mars reappears on our road to the Big Bang. We have already met Percival Lowell and his belief that there were canals on the surface of Mars and this meant to him that there must be a civilisation on that planet. In order to further his investigations he set up the 'Lowell Observatory' on what is now "Mars Hill', Flagstaff, Arizona. The primary aim of the observatory was to look for life on Mars and in order to pursue this aim they borrowed two telescopes, one from the Harvard observatory and a second from the manufacturer John Brashear, in Pittsburgh. One year later, their own twenty-four inch refracting telescope was ready and so the time came to return the two telescopes that they had borrowed to their original owners.

Being naturally grateful for the loan of the instruments they decided to give the lenses a good clean before returning the telescopes and, Douglass, the director of the observatory, is alleged to have suggested that alcohol would be the best cleaning agent. An underling was sent to the local town store for some 'good alcohol' with which to clean the optics. Somewhere along the line a misunderstanding occurred and instead of bringing back some 'good alcohol' as requested he brought back 'wood alcohol'. It appears that no one at the observatory bothered to check the label on the bottle and they proceeded to clean the lenses with this - with the unfortunate result that it damaged one of the lenses on the telescope borrowed from Brashear, the manufacturer. Now this left Lowell in an extremely embarrassing situation and the only thing left for him to do was to return the telescope, own up to the damage and offer to pay for it to be repaired. The bill came to some four hundred dollars and so Lowell immediately wrote a cheque and sent it to John Brashear, the owner of the firm in Pittsburgh who immediately ripped the cheque in two and refused any payment.

Being a man of honour (even though he believed in Martians) and still being embarrassed by the whole situation Lowell came to a compromise. If he was not allowed to pay for the damage, he would order a new instrument from Brashear for use in the observatory and so made the 'guilt' purchase of a new spectrograph. He then fired the observatory director, Douglass - after all, it was he who had recommended the alcohol. Thus it was that the instrument that would be the first to measure the expansion of the universe would come into being.

Lowell persuaded Vesto Slipher to come and work at the observatory. Slipher was born in Indiana and had graduated from the University of Indiana where he had studied astronomy. Slipher was set the task of using the new spectrograph to study the atmospheres of the planets and their speeds of rotation. In 1908, as part of the drive for the Lowell observatory to find intelligent Martians on Mars, Slipher reported that he had used the spectrograph and, contrary to others, had found the existence of water vapour in the Martian atmosphere, a necessary condition (but not proof) of life on Mars. To measure the speed of rotation of planets he used the Doppler Effect.

Planets reflect the light from the Sun and when a planet rotates, one side is turning towards the Earth and light reflected by it should be blueshifted whilst the other side is rotating away from Earth and so light reflected from this side will be redshifted.

Whilst the primary aim of the astronomers and the observatory itself was to look for evidence of life on Mars, this work cannot take place the whole year round since Mars is only visible at certain times from the Earth. During Martian 'down times' the astronomers were able to pursue other research projects, and Slipher took the opportunity to turn his attention to some 'fuzzy' white patches called nebula. A nebula is a cloud of gas and dust but at this time, no one knew the difference between nebulae and galaxies, they all looked the same to them. There was some controversy growing as to what they were and it was Hubble and his distance measurements that finally solved the problem – but this is still in the distant future and this work had yet to be done.

In 1912, the same year that Leavitt published her work on the variable stars in the Magellan clouds and the Titanic sank, Slipher turned his spectrograph onto the Andromeda galaxy, applied the Doppler principle and was stunned to find that the galaxy was moving towards us at the staggering rate of three hundred kilometres per second (just under seven hundred thousand miles an hour!) but this pales to a mere fifty kilometres per second once the speed of our solar system is taken into account as we are moving towards Andromeda due to the rotation of our own galaxy, the Milky Way. Slipher continued this work with other galaxies and by 1914 he had measured the velocities of thirteen galaxies and of these only two were moving towards us (blueshifted) the rest were moving away from us at great speeds (redshifted). The reason it took two years to gather these results was partly because of Slipher's methodical work ethic (he would only publish when he was absolutely sure) and partly because of the difficulties involved. Very little light arrives from these distant galaxies and very long exposures had to be used to expose the photographic plates. Each exposure had to be made over a series of several nights during which time, Slipher had to continually 'tweak' the direction of the telescope so that the galaxy remained in target, and Slipher would perform all this whilst wearing a suit and tie.

Slipher presented his results in a paper addressed to the annual meeting of the "Astronomical Society of America" and at the end of his presentation was rewarded by a standing ovation from all those present. Never before had such speeds been recorded anywhere. They were far in excess of anything measured before - and add to this the fact that nearly all of the galaxies were moving away from us at these tremendous speeds, it was enough to bring any astronomer to his feet. Even without knowing the distances this was enough for astronomers to postulate that with speeds like these, galaxies must be outside our own Milky Way and that the Milky Way was not the whole Universe but only a small part of it. But science does not work on thoughts, it needs firm experimental evidence and that would have to wait for the young Hubble to arrive.

When Lowell died, he left the observatory one million dollars to continue the work and made Slipher, the director of the Lowell observatory. He went on to measure the radial velocities of over forty galaxies whilst astronomers in the rest of the world only managed a paltry four, and apart from the original two, every galaxy was found to be moving away from the Earth at a great velocity. No mean feat in itself as Lowell's wife had objected to the will and went to court with the result that most of the million dollars left for research went in legal fees. On the other hand, Slipher's brother worked at the Lowell observatory and, as late as nineteen sixty-one was still insisting that there was life on Mars.

As luck would have it, sitting in the audience, listening to every word Slipher said and joining in with the ovation at the end was one Edwin Hubble, on his way to take up a place at the University of Chicago to study for his PhD.

Chapter 9. Bangers and mash.

Albert Einstein was born in Germany, in 1879. His father owned both a tavern and an electrical business but he went broke shortly after young Albert was born and so they moved to Munich, where his father set up yet another electrical business. Albert was a strange child in that he did not speak until he was three years old and even then said very little; what he did say, he said very slowly. Albert had no respect for authority and irritated his teachers so much that, at one time he was expelled from school. This meant that he had to gain much of his knowledge from relatives and friends. He eventually qualified to be a teacher of Mathematics and Physics but was unable to get a job as his teachers refused to give him a good reference. As a result, Albert took a job at the Swiss Patent office. Whilst there, he continued to ponder the meaning of the Universe and in doing so, ignored his wife so much that it more than likely led to their divorce. He published his theories and, as his work became recognised, he was able to rise to the posts of University lecturer, then Professor and, in 1921, he was awarded the Nobel Prize for Physics "for his services to Theoretical Physics, and especially for his discovery of the law of the photoelectric effect". Here, he showed that light behaved like a particle, consisting of photons, each with energy dependent upon their frequency. It must be remembered that a great deal of his early work was done in his spare time whilst working at the Patent Office. Albert was a scruffy individual with unkempt hair and ragged clothes but he was also a pacifist. He proposed that the entire World should have only one government and, during World War One, objected to the use of science to develop new weapons, which is strange because it was Einstein who, at a later date, wrote to President Roosevelt in the USA, proposing the idea of a nuclear bomb! He suggested the idea of a nuclear bomb, as he thought the Germans were developing one too and did not want the Americans to be left behind! Unfortunately his information was wrong, as it appears the German's had changed their minds and decided against the idea after all. Einstein was Jewish and in 1952 he was asked to become the President of Israel, a job he turned down on the basis that he was too old and didn't have enough experience (it hasn't put anyone else off has it?). Albert continued his work at Princeton University, spending the last thirty years of his life trying, but failing, to develop a unified field theory. On his deathbed, he made a supreme effort to leave to the World his last few words of wisdom. Had he finally solved the puzzle of the last thirty years? Had he come up with a new theory to revolutionise the scientific world? Was it a way to prevent future wars? Unfortunately, we will never know, as he spoke the words in German and there was only one nurse with him at the time he died, and she only spoke English! What do you think Einstein's last words would have been?

After Einstein's death, his brain was preserved for prosperity and examined to try to find out what made the genius a 'genius'. There have been three studies, the latest in 1999, and this showed that his brain was a little wider than 'normal' brains and had a 'groove' on the surface 'missing' which every 'normal' person seemed to have. Naturally, this proves nothing as it may just be a fluke or the effects of the preserving fluid – we need a lot more brains from geniuses to see if they are all missing this same 'groove' before we can tell if it means anything - any volunteers? There are lots of bits and pieces of famous and clever people stored around the world. Along with Einstein's brain which was kept by the surgeon who removed it, there is a finger cut from Galileo which is pickled in a jar in Florence, Italy and at Trinity College Cambridge we have a lock of Newton's hair - which leads us to an interesting thought. We can already clone sheep and some say that we should be able to clone hairy mammoths from the remains of animals that died long ago and are found in the frozen wastes of Siberia. So if all this is possible, why don't we clone these brilliant people and form Einstein's 'one government' to rule the World? If we cloned Newton, Galileo and Einstein we could create his 'World Council' of brilliant people who would guide us on our way and solve the problems of all and sundry. Edwin Hubble clearly decided not to be a part of this council since, on his death, his wife scattered his ashes and did not tell anyone where they where - so not only can we not clone him for our council, we cannot even find his grave! We could also go one step further and throw in a few religious people for good measure. January first is "The Feast of the Circumcision of Jesus Christ" and quite a few churches say that they still have Jesus' foreskin so, if we determined which one it was, we could clone Jesus and have him on our council too. Perhaps not, let's get back to the story!

Einstein had formulated his theory of general relativity and this included a mathematical model of the Universe. In the general theory of relativity, the geometry of space is not 'flat' like the top of a pool table, but 'curved'. In 'flat' space, if one rolls a ball across a pool table the ball will travel in what everyone will agree is a 'dead' straight line. This is not the case in Einstein's 'curved' space as here any masses or chunks of energy will 'bend' both space and time so regardless of where we are in our relativistic universe, our surfaces are no longer flat and our watches will not run at the same speed. The 'oft quoted' example, to explain the curvature of space is that instead of playing our game of pool on a completely flat table, in general relativity, we must set about playing pool on a trampoline. Here, masses such as the other pool balls in play, cause the surface to 'sag' or 'curve'. When we roll the ball across the surface, whilst the ball itself thinks it is following a straight line, everyone watching from the sides will say that it follows anything but a straight path. They see it going in and out of the dents formed by the other balls, which are stretching the elastic material from which the surface is made. As to whether it is straight or not depends upon one's 'frame of reference' – something like a drunken man walking home. He sees himself as walking in a straight line but onlookers see him staggering from pillar to post! Einstein took his general theory and applied it to the universe hoping that it would not only describe, mathematically, the whole universe, but he was also hoping that there would be one and only one solution to his equations. This would show that our universe was the only one possible and that there were no others! At that time everyone believed that the Universe was static, infinite and had been and always would be, around here forever. That is, it went on forever and ever, never becoming bigger or smaller, an unchanging Universe, one that we could rely upon to be the same – always. This was not enough for Einstein, as he wanted some quantitative details of the universe. That is, do the mathematics and put some numbers in.

Unfortunately, Einstein's equations predicted that a static Universe was not possible; His equations said that the Universe had to be either expanding or contracting – it could not remain static as Einstein and everyone else fervently believed. For this reason and this reason alone, Einstein introduced a 'fiddle factor', called the 'cosmological constant', into his equations because he did not believe what the equations were telling him and he wanted a solution that would make a static Universe possible.

Willem de Sitter was a mathematician born in 1872 in Sneek, a town in the Netherlands. He became a Professor of astronomy at Leiden in 1908 and went on to become the Director of the Leiden Observatory in 1919. He was also interested in applying Einstein's equations to the universe and published his results just after Einstein published his. In de Sitter's universe, the universe was static, but the small problem was, there wasn't any matter in it. That is, it was completely empty! Space was static but there was nothing for us to stand on. No Earth, no Sun, no galaxies or anything. This might seem a small difficulty to you, when one thinks that this is a theory to describe the universe we live in, but remember, we are talking about theoreticians here so it doesn't matter! To try to rectify the situation, de Sitter tried to put some tiny pieces of matter into his model but as soon as he did that the whole thing started to expand. General Relativity was telling de Sitter that a universe with mass had to expand. The same theory also predicted a redshift in light travelling from one mass to another.

Meanwhile, the Russians were not going to be left out of this relatively new aspect of science. Alexander Friedmann was born in St Petersburg, Russia in 1888, and in his short life, he lived through some of the most troubled times in that country. Friedmann was a great mathematician but the First World War interrupted his university studies when Russia found itself fighting both Austria and Germany. Friedmann volunteered to be a bomber pilot and in his spare time between sorties, he set his mind to the mathematics of bomb trajectories. To see if he had obtained the correct solutions to the problem he would go out the next day and drop a few bombs on the Austrians, who readily confirmed that he had got his sums correct!

His mathematics project was disrupted by the Russian revolution and the ensuing civil war between the red and white armies but he came through all of this, ending up back in present day St Petersburg. It was here that he took an interest in Einstein's General Theory of Relativity and Friedmann realised that it was space itself that was expanding, carrying the galaxies along with it. He wrote an article on "The Curvature of Space" and published it in a scientific journal in 1922. This article invoked a quick response from Einstein who had a letter published in a later edition of the same journal, stating that Friedmann was wrong! However, Friedmann sent his calculations to Einstein for him to check and Einstein eventually went over them and gracefully sent another letter to be published stating that it was he, Einstein, who had made the error and Friedmann's calculations were indeed correct.

Thus it was that the curvature of space was thrust upon us. Friedmann himself died of typhoid, in 1925, at the age of thirty-seven. Some say that he died from 'catching cold' from one of his balloon ascents. Friedmann was a meteorologist who made balloon ascents to gather data and even held the World record for the highest ascent. However, having a Russian wife myself, I know that to a Russian, "catching cold" is a euphemism used for any illness ranging from a headache to the Black Death and could quite easily include typhoid!

Meanwhile, a Belgian Catholic Priest, Georges Lemaitre, had been going over Einstein's equations and Friedmann's work and became convinced that the Universe was expanding. Lemaitre was born in 1894, trained as a civil engineer and took holy orders after the First World War. He studied astrophysics at both the University of Cambridge, UK, and at the Massachusetts Institute of Technology before becoming Professor of Astrophysics at the University of Louvain.

Lemaitre removed Einstein's fiddle factor, and in one fell swoop had consolidated science and religion in that they both had a point of creation. An infinite Universe has been here forever and so it is not created. This goes against religious beliefs but a Universe expanding out from one specific point in space is a universe that has been created. Christianity would not be the only faith to benefit from an expanding universe as I am told that the Holy Qur'aan also includes a reference to an expanding universe. Chapter ath-Thaariyaat of the Holy Qur'aan translates as:

"I built the heaven with power and it is I, who am expanding it" Qur'aan, 51:47

Lemaitre had heard of the radial velocities of the galaxies whilst studying in America and he decided that both the general theory of relativity and experimental results showed that the Universe was expanding. In Lemaitre's theory, the Universe had matter within it and was also expanding. Einstein's expanding universe that contained matter and de Sitter's static but empty universe could be extracted as 'special cases' from Lemaitre's theory. He published his results in 1927 but, since he was totally ignored, he went to see the great man, Einstein, himself. Lemaitre met Einstein in Brussels and after discussing Lemaitre's results, Einstein told him that his mathematics was fine but his Physics was ... well, not too good. In other words, Einstein told him that he was wrong! Einstein does not seem to have learned much from the insults of Friedmann five years earlier!

Cosmologists tend to prefer Friedmann's 'curvature of space - time' as the driving force behind the expansion of the Universe rather than the Doppler Effect (even though they may still refer to the Doppler Effect in public) as galaxies and Heavenly bodies have been found with speeds of the order of the speed of light. With the Doppler Effect, one would expect these bodies to have been distorted as they were accelerated to these huge speeds. Since they show no apparent distortion and are, to all intents and purposes, 'normal' cosmologists prefer to say that it is space itself that is stretching – and carrying the galaxies along with it, rather than the galaxies themselves moving along.

The scene is now set for the merging of two separate areas of astrophysics and, for a major breakthrough. The discovery of

Cepheid variable stars had enabled Hubble to use them to measure the distance to the Andromeda galaxy. Slipher had cast aside his search for Martians and had measured the radial velocities of galaxies and so it seemed a natural progression to see if there was a link. Several scientists had already proposed that there was a link between distance and redshift because they had realised that the dimmer a galaxy appeared, the greater the redshift it had. In astronomy, 'dimmer' usually means further away, but it is one thing to suspect a relationship between distance and redshift and quite another to prove it. What science needs is proof and 'he who does the work gets the credit' for the discovery.

Slipher had measured the radial velocity of several galaxies but his telescope only had a twenty-four inch objective lens. In astronomy size really does matter and Slipher's twenty four inches just weren't enough to bring the most distant heavenly bodies within his reach. A telescope is classified by the diameter of the objective lens or mirror that collects the light. The wider the lens or mirror then the more light it will collect and thus enable one to see the dimmer and hence more distant galaxies. Increasing the magnification is easy; making large diameter lenses and mirrors that do not distort the light is expensive and difficult.

Meanwhile, at the top of Mount Wilson was the one hundred inch Hooker telescope that was then the World's largest telescope. The driving force behind the construction of this and several other large telescopes was George Ellery Hale. He had persuaded J. D. Hooker, an industrialist from Los Angeles, along with the Carnegie Institution to build this huge telescope nearly six thousand feet above sea level at the top of Mount Wilson, California. The telescope marks a significant development, not only in astronomy but in engineering as well. The problem is that the Earth rotates once every twenty-four hours and this makes the stars and galaxies appear to move in the opposite direction. To allow for this, the whole telescope has to turn easily and smoothly in the opposite direction to the Earth, thus enabling it to track and photograph distant galaxies. A telescope of this size is heavy; the mirror alone weighs around four tons, and so to make it moveable is not an easy task. The whole telescope, all ninety tons of moving parts (the steelwork was constructed by a shipbuilding company), floats on large vats of mercury, enabling the whole telescope to be moved at the touch of a hand (well, a big touch and provided it is 'touched' at the end of the tube – 'give me a lever long enough and I will move the Earth?). The mirror was cast in France from the same glass as that used to make wine bottles (probably including a few 'empties') and then sent to the US to be ground into the correct shape. Casting the disc that would eventually be ground into the mirror was not easy. It was not a case of 'drink the wine, collect the empties, melt the glass, pour it into a mould and that's it done'. No, the glass had to be cooled very slowly to keep out any imperfections such as air bubbles and even then there were still a few air bubbles left in the final disc. Eventually, the technical problems were overcome and the telescope built. It is now listed as an 'International Historic Mechanical Engineering Landmark.'

Edwin Hubble was born in Missouri in 1888. His father was a lawyer working in the insurance business and, as a result of his father's work, the Hubble family left Missouri to take up residence in Chicago. At school, Edwin was an able student who could get by on just his natural ability alone. He did not need to work too hard to succeed and so he didn't over exert himself. Despite this, he excelled both in his examination results and in sports with the result that he was rewarded with a scholarship to the University of Chicago to study Law.

When talking of Hubble, it is difficult to separate the truth from the legend. In his lifetime, he was both a great astronomer who was revered by his peers and a figure constantly in the public eye, mixing with scientists, film stars and dignitaries alike. It must also be said that unlike his astronomical calculations, many of the stories about him, and even told by him, just do not add up; they are not lies but 'exaggerated' shall we say.

At Chicago University he excelled both at his studies and in athletics and, since his father would not let him play football, he took up boxing. He won a coveted Rhodes scholarship, which gave him the opportunity to spend three years studying law at Oxford University, eventually graduating with a degree in Spanish (yes that really is Spanish). In England, not surprisingly, Hubble came to love everything English, the way they dressed, behaved and talked and this was to remain with him for the rest of his life. From this time on, he dressed like an Englishman with cape and cane, and spoke like an Englishman, but with an affected English accent. His father had died whilst he was in Oxford and, since Hubble needed a job, he spent one year teaching at the local high school before securing a place in the graduate school of the University of Chicago. It was there that Hubble started his career in astronomy and took his doctorate, performing his observations at the University's Yerkes observatory. One day, whilst on his way to the Yerkes observatory, he stopped off to listen to Slipher's presentation on the radial velocities of galaxies little knowing that this would lead to one of his greatest achievements.

Hubble finished his doctorate thesis in a rush and enlisted in the army, as by then war had broken out, but first he obtained a promise of a job at Mount Wilson after the war had finished. During the war years, Hubble was based yet again in England and is alleged to have seen action in France, being promoted to the rank of Major (with the result that even more affectations would be added to the archetypal English ones already adopted).

After the war he returned to Mount Wilson where he devised a system to classify galaxies into different types. He showed that there was more to the Universe than the Milky Way by measuring the distance to the Andromeda galaxy, a feat which most astronomers would be happy with even if it was their only achievement in their entire lifetime. Hubble, assisted by Humason, set about repeating Slipher's redshift measurements to find the radial velocity of galaxies but he also used Cepheid variables to measure the distance to these same galaxies. The result was that they found a pattern. Galaxies twice as far away had twice the redshift, whilst galaxies three times as far away had three times the redshift. Since in those days, redshifts were associated with velocities, this was interpreted as: the radial

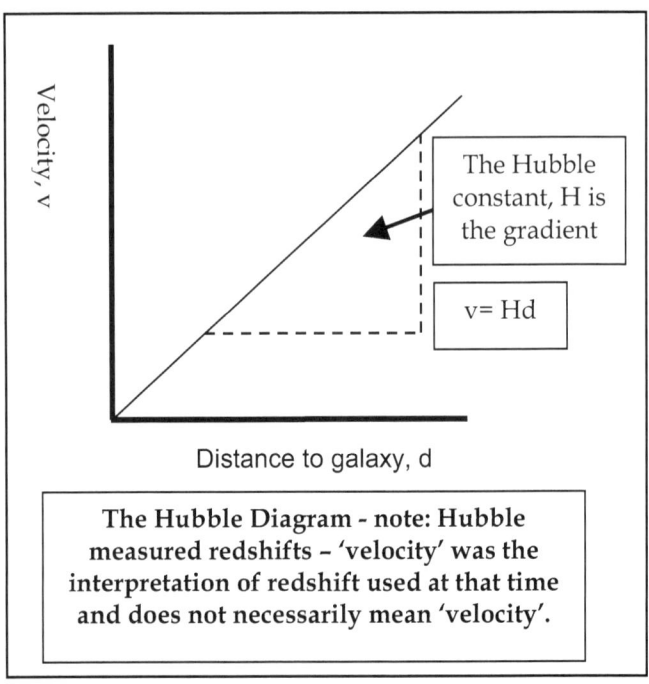

The Hubble Diagram - note: Hubble measured redshifts – 'velocity' was the interpretation of redshift used at that time and does not necessarily mean 'velocity'.

velocity of the galaxy (v), is directly proportional to how far away the galaxy is (d) and the constant of proportionality is now known as the Hubble Constant and given the symbol, H.

$$v = Hd$$

Hubble displayed the results as a graph of 'velocity' against time and decided that the points formed a straight line through the origin, the gradient of which is the Hubble constant itself (since the errors were large he needed a good imagination to come up with this pattern).

When Hubble published his paper on the distance velocity relationship, Lemaitre's original paper was published again in England resulting in even more support for his claims that the general theory predicted expansion. However ideas that have been hammered into one's brain are always hard to expunge and scientists were still reluctant to accept that the entire Universe had been created at a single point and at a certain time. Lemaitre pictured the creation of the Universe as a giant firework going off, exploding and throwing 'glowing embers' out in all directions. The 'glowing embers' pictured by Lemaitre, were the galaxies moving away from each other and he felt that by working backwards we should be able to show that there was a point of creation. That is, imagine taking a movie of a firework exploding in the sky. When the film is run backwards we see the embers receding until they coagulate at one point. This point marks the place where the Universe was created.

This gave great importance to the Hubble Constant, H itself as this gave a way of determining how long ago this firework display was – in other words, he who knows the Hubble Constant knows the age of the Universe!

Hubble had shown that:

$$v = Hd$$

Where v is the velocity of a particular galaxy and d is its distance away from Earth. At first glance this seems just too good to be true. Why is it that galaxies twice as far away from Earth should just happen to be going twice as fast and so on? Until that is, we realise that if everything started off in this 'giant firework' then everything started moving outwards at the same time. Pieces of matter that were thrown out from the initial 'Bang' twice as fast would have travelled twice as far in the same time. Pieces of matter that were thrown out from the initial 'Bang' three times as fast would have travelled three times as far in the same time - thus giving us the Hubble Law. If the age of the Universe is 't' then all matter must all have started out from the same point a time 't' ago. This time 't' is also how long it takes galaxies to move the distance 'd' away from us here on Earth. At the point of 'creation' all matter was side by side but has been moving apart ever since the Big Bang. We measure how fast a distant galaxy is moving away from 'us' and so the distance between that galaxy and Earth must be how far that galaxy has moved away from Earth during the lifetime of the Universe.

$$(\text{Velocity, } v) = (\text{distance, } d)/(\text{time,}$$

Or

$$v = d/t$$

Since the velocity, v and the distance d are the same in both equations we can put them equal to give:

$$Hd = d/t$$

Dividing both sides by d and re-arranging the equation gives:

$$t = 1/H$$

Therefore, the reciprocal of the Hubble constant gives us the age of the Universe. This made the Hubble Constant a very important number indeed.

Einstein and de Sitter got together and, in 1932, published a joint paper setting out the Einstein – de Sitter model of the Universe. In the following year, 1933, Lemaitre and Einstein both found themselves in California and a meeting of the 'big three' Einstein, Hubble and Lemaitre, was inevitable. At this meeting, Lemaitre went over Hubble's results and Einstein's equations and convinced them both that the Universe had indeed started in a 'single quantum' at a point of creation. Einstein went on to describe the introduction of his cosmological constant, a constant he had introduced to make the universe 'static', as the biggest blunder of his life (though it was not long before Lemaitre had to reintroduce the constant to make his sums fit observed expansion rates. He later removed it as this turned out to be a further blunder. It was the measurements that were incorrect and not the theory). Hubble too, we are told, was also convinced at this meeting that the universe was expanding and thus an expanding universe was born. It must also be said that Hubble decided against an expanding Universe a few years later and Einstein's 'cosmological constant', has more recently had to be reintroduced to make theory 'fit' present day experimental results. In fact in cosmology, Einstein's 'cosmological constant' has been very much like the song, "In, out, in, out, shake it all about".

Let us just pause for a moment to consider this great meeting of minds; a meeting where the origins of the Universe were finally thrashed out. We have Hubble, a confirmed anglophile, dressed in his plus fours (knickers to Americans) cape and cane, acting like he was still a major in the army and speaking with the most affected of British accents. We have Lemaitre, a Catholic priest with dog collar and sometime wearer of his badge of office, the frock. Finally, there was Einstein, with long scruffy hair and wearing no socks, in the belief that since his big toe would always come through the socks it was best not to wear them. Would you buy a second hand car from these people let alone a new Universe? Well lots of people did and the expanding Universe came into being. Until 'along came Fritz', and by all accounts he was no ordinary guy!

Fritz Zwicky was born in Bulgaria in 1898 and, since his parents were Swiss he was sent to be educated in Switzerland. He eventually moved to California to work with Millikan (who had already discovered the electron) at the California Institute of Technology (Caltech). Zwicky took an interest in astrophysics and it is mainly in astrophysics that he made a name for himself both as a genius and as a loud mouth! It is difficult to find a reference to Zwicky without some mention of his character. Mercurial, insufferable, unusual or just plain crazed are all adjectives used by colleagues to describe him, but he was responsible for several of our present ideas in astrophysics. He took photographs of large areas of night sky and showed that galaxies were not evenly spread out but were found in groups called clusters and super clusters. He (along with Walter Baade) gave us neutron stars where a star collapses so much, that the electrons are forced into the protons in the nucleus to give nothing but neutrons. He gave the name 'supernova' to the world and was the first to suggest that this is where all cosmic rays originate. He was the first to see that galaxies must contain much more matter than we can see through a telescope. By measuring the brightness of a galaxy, he estimated how much matter it contained but, when he calculated how much mass was needed to hold these same galaxies together he found that their mass had to be up to four hundred times the mass that could be seen. In doing this he started the search for what we now call 'dark matter' or the 'missing' mass. In 1937, he put forward the idea of gravitational lensing where galaxies 'bend the space' around them in such a way that they magnify objects behind them.

However, read any book and the author will tell you how he had a persecution complex and thought everyone was against him or was stealing his ideas. Robert Kirshner, in his book, "The Extravagant Universe" gives a very lucid description of Zwicky. He allegedly called everyone 'bastards' or, if you were particularly lucky a 'spherical bastard' meaning that you were a bastard from every point of view. When he did not know the name of one of his students, we are told that he would ask them 'who the hell are you?' and the effect on the students would be amplified by Zwicky's gnomish appearance; short stature with pointed eyebrows.

Hubble would hold meetings in the front room of his home on a regular basis and it was there that astronomical problems or new ideas would be thrashed out with the aid of chalk and a black board. At one of these meetings the point under discussion was the explanation of the redshifted spectral lines and whether this was due to Lemaitre's expansion theory. Fritz Zwicky was present at this meeting and, always one to come up with an original idea, put the idea forward of 'Tired Light'. Zwicky proposed that as photons of light from distant galaxies travel towards us, they are 'held back' by the gravitational forces and thus lose energy. The velocity of light is a universal constant and thus light cannot slow down as normally happens when something loses energy. With light, the energy of a photon is related to its frequency and so if it loses energy the speed will stay the same but the frequency will reduce and as a consequence of this, the wavelength will increase. That is, it will be redshifted. Furthermore, Zwicky proposed that the further the photon travelled, the greater the effect of gravity holding the photon back would be and so the greater the redshift. In this way, the experimental evidence of redshift could be explained. He also claimed that the redshift-distance data was not a straight line as Hubble claimed but was exponential in nature.

$$z = e^{Hd/c} - 1$$

This is an important relationship and we will meet it again later.

Zwicky was surprised to find that no one took his idea seriously apart from Hubble that is, who, whilst not agreeing with Zwicky's 'Tired Light' theory, from that day on, refused to call the redshifts 'velocities' any more. This was on the basis that it was redshift that was measured by experiment. As to what these redshifts represented, a velocity or energy loss, was open to interpretation. That afternoon's meeting marked the birth of 'Tired Light'.

Zwicky's Tired Light theory did not catch on as there was no known mechanism by which the light could lose energy in the way that Zwicky had proposed. The explanation of redshift continued to be that it was a result of an expansion of the Universe. This too had its problems since at that time, no one understood the atom, let alone what it could possibly have been that had 'exploded in the first place (the neutron had yet to be discovered). Not only this, but an explanation had to be found as to how the various elements had managed to form. However, help was at hand in the form of a young Ukrainian and his wife, paddling across the Black Sea in an attempt to escape from Russia and defect to Turkey.

George Gamow was born in Odessa in what is now known as the Ukraine and he completed his university studies in St. Petersburg. On receiving his PhD, he worked in research institutions in Cambridge, UK, and then in Copenhagen but, in 1931, Gamow received a missive from the Communist state of Russia demanding his recall so that he could be placed in charge of research at the Academy of Science in St. Petersburg (Leningrad at that time). Communist Russia did not suit Gamow and with the Nazi shadow starting to fall over Germany at the time, he decided to defect to the West. This was made harder by the fact that he had married and so there were two of them in need of an escape route.

Gamow and his wife, Luybov made several escape attempts but according to Gamow, the most daring was probably when he and Luybov took a small canoe and set off with a small amount of food, but plenty of alcohol, to cross the Black sea in an attempt to reach Turkey. They paddled for one and a half days and covered just less than two hundred miles before the weather changed and the wind began to howl. The winds became so strong that it blew them back faster than they could paddle forwards. They were eventually blown back to Russian shores and the tired and exhausted pair were taken to hospital. Of course the problem then was how to explain their ordeal to the communist authorities as trying to defect was an offence and deeply frowned upon. Gamow managed to persuade the Communist powers that they had been testing their 'new boat' when the bad weather had driven them out to sea and it was only by the greatest of efforts that they had managed to return to good old Russia.

He must have been a good liar because in 1933, after another failed attempt to escape, he was selected by the Russian authorities to represent the Soviet Union at a conference on theoretical Physics to be held in Brussels. Gamow contrived a meeting with Stalin's number two, Vyacheslav Molotov (for cocktails, no doubt) where he persuaded Molotov to allow him to take his wife Luybov along with him as his secretary. Well, at least it would save the Soviet State money, as they only needed one hotel room. Off they went and never came back - ending up in the USA where they stayed for the rest of their lives, but divorcing in 1956.

Gamow's first big contribution to science was in radioactivity. Whilst still young, he put forward the explanation of alpha decay, explaining how some isotopes decayed very rapidly whilst others took thousands of years. It was this theory that brought him to the attention of the World and proved to be the ticket to take him to meet both Ernest Rutherford in Cambridge and Neils Bohr in Copenhagen.

Like a child with a new toy who takes it apart only to put it back together again, Gamow had looked at the Physics of atoms falling apart, so he started to look at the possibilities of putting them back together again in a process known as nucleosynthesis. At Cambridge, Gamow had played a major role by showing the feasibility of an idea by Cockcroft and Walton to accelerate protons and wham them into the nuclei of atoms to produce 'new' atoms. Whilst living in the USA he had been at Los Alamos, working on the development of the nuclear bomb. Consequently, he had the necessary background to explain the formation of the various elements in the Universe.

A research student, Ralph Alpher, joined Gamow and between them they made the first attempt at explaining how it all happened. They proposed that, in the beginning, the Universe started with a hot Big Bang, made up principally of neutrons. A neutron is a gregarious particle – it likes to be amongst friends! Whilst it is in the nucleus of an atom being 'chummy' with the other particles, it is a fairly stable particle and does not decay, but when it is all by itself, it just goes to pieces! It breaks up into a proton and an electron fairly rapidly. In fact if there are a group of separate neutrons, then half of them will have broken up in just over ten minutes. So as the Universe expands, the neutrons spread out and decay into protons and electrons. Eventually (but not yet) these two particles will recombine to form an atom of Hydrogen and, since this is a very likely process, we would expect most of the Universe to be made up of Hydrogen, and it is. Protons are positively charged and this makes them repulsive little things, so a collision where a proton bumps into another proton and sticks to it, is unlikely, but a neutron can bump into a proton (the nucleus of a Hydrogen atom) to make an isotope of Hydrogen known as Deuterium. If a second neutron bumps into the Deuterium nucleus it will produce another isotope of Hydrogen, Tritium. However, Tritium is very unstable and one of the neutrons will decay into a proton, throwing out an electron. This gives us three particles in our nucleus, two protons and one neutron, so when a third neutron whams into this nucleus, we get the nucleus of a Helium atom. The atom will later pick up the extra electrons from the particles flying about in the dense 'cosmic soup' of particles to form atoms. The next step is where it all goes wrong because we do not get five particles in the nucleus of an atom. To reach the next and heavier elements we would need one very big coincidence to happen. We would need two particles to hit our nucleus at the very same time and stick to it! This might have seemed probable in our dense 'pea souper' of a Universe but remember, our Universe is expanding and the density is getting less all the time and so we are expecting this already unlikely situation to occur when the chances of particles colliding are even more remote. In its defence, Hydrogen and Helium do form about 99% of the Universe and these heavier elements are fairly scarce on the grand scale of things but, as we shall see later, the heavier elements are formed elsewhere. But that is the way it is and how Gamow and Alpher explained the formation of the lighter elements in the Universe.

They published their results on April first, 1948, and Gamow decided upon a little joke. Even though Hans Bethe is said not to have had anything to do with it, Gamow included his name on the paper so that the list of authors read,"Alpher, Bethe, Gamow" the first three letters in the Greek alphabet.

This paper is generally accepted to be the birth of the 'Big Bang' and so the Big Bang theory first saw the light of day as a joke published on April fool's day! Does that tell us anything?

Not all scientists were happy about the Big Bang theory and Hoyle, Gold and Hermann set out to disprove the theory - but in doing so, they ended up by working out the entire details! Gamow's double collisions to make the heavier particles was just too unlikely and it was Fred Hoyle, along with Willy Fowler and Geoffrey and Margaret Burbridge, who corrected the matter and showed that the heavier elements are produced when a star collapses, with the heavier elements distributed throughout the Universe when a star explodes in a tremendous explosion, known as a 'supernova'.

There was a problem with Gamow's explanation of how the lighter elements were formed in the early Big Bang, quite a big problem in fact. With all the protons bumping into neutrons to form deuterium and then bumping into further neutrons and protons to form Helium, there was no reason for this process to stop, and it wouldn't. All of our particles would end up as Helium and there would be no Hydrogen left in the Universe. We know that this is completely untrue as most of our Universe (75% by mass) is made up of Hydrogen. Gamow needed a bright idea and he came up with one in the form of gamma rays. When the first proton and neutron whammed into each other to form deuterium, if a high-energy gamma ray came along and bumped into the pair, it could split them up and thus prevent the process continuing to Helium. The ratio of the densities of gamma rays to protons and neutrons is crucial as is the energy of the gamma rays at a particular time. Too many gamma rays, and all the deuterium nuclei will be split up and we will not move on to Helium, leaving us with a Universe made entirely of Hydrogen. Too few gamma rays and all the protons and neutrons will go on to form Helium, giving us a Universe depleted of Hydrogen.

Gamow *'chose'* a value for the ratio of the density of photons to baryons ('baryon' is a general category of particles including protons and neutrons) and he chose one that would give him the 'correct' ratio of Hydrogen to Helium. He then realised that, as the universe expanded there would have to be a 'cut off' point where the gamma rays could no longer break up the deuterium. This cut off point would be when the Universe was a few minutes old, when around three quarters of the original neutrons had decayed into protons and electrons. In the first few minutes, the gamma rays could split up the deuterium and prevent it from going on to form Helium. After this time, the universe had expanded and stretched the gamma rays so much that they no longer had enough energy to break up the deuterium. The nucleosynthesis process could go on unheeded, resulting in the correct ratio of Hydrogen to Helium. That meant that the energy of the gamma rays at the time the Universe was just a few minutes old, was also critical as it not only determined the ratio of the light elements but it also set the constraints upon the Cosmic Microwave Background (CMB) radiation – the Black Body radiation that is all around us till this day.

A perfect 'Black Body' is one that absorbs all the radiation falling upon it, reflecting none. Black Bodies are not only the best absorbers of radiation but they are also the best emitters of radiation (just as we learn at school – solar panels are painted matt black as this absorbs most of the radiation that falls upon it and so heats our swimming pools, and yet car radiators are also painted matt black because this colour emits most radiation and keeps your car engine cool). Perhaps the best way of reproducing a black body is to have a can with a small hole in it. Any radiation entering the can through the small hole bounces around the inside of the can until it is eventually absorbed - since it is unlikely that it will find an escape route. Since any light entering the can from the outside is absorbed, never to be seen again by the outside world, the can is a perfect absorber of radiation and hence a 'Black Body'.

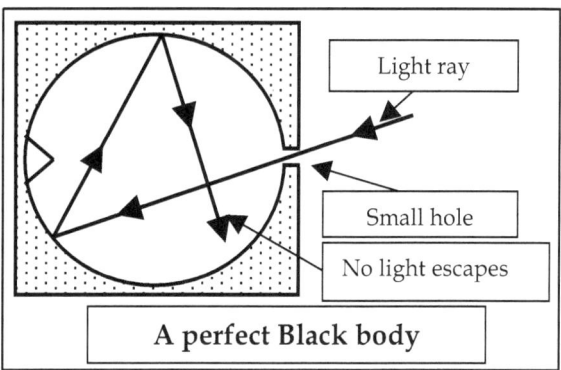

A perfect Black body

If we heat the can then it will radiate energy and, since any radiation emitted will have bounced around the inside many, many times before finding the hole through which to exit, the spectrum produced will be independent of any variations in the inside surface of the can. The spectrum of the emitted radiation will depend only on the temperature of the Black Body radiator.

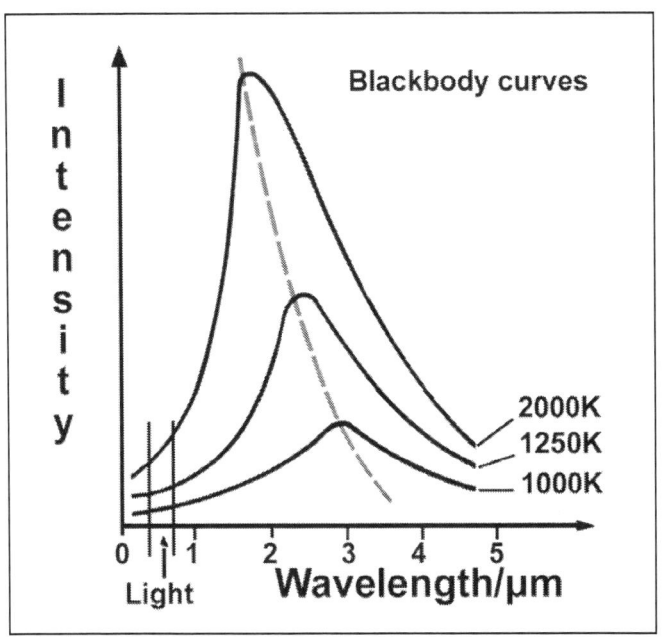

Gamma rays are 'trapped' inside the expanding Universe and so they too act in the form of 'Black Body' radiation. We can relate the energy of the gamma rays to a temperature of more than one billion Kelvin during the first few minutes of the Universe. The gamma rays are thus able to split up the deuterium nuclei. After this time, space had expanded and stretched the wavelength of the photons so much that the frequency and hence energy of the gamma rays had reduced - meaning that the temperature had fallen below one billion Kelvin and the gammas were no longer able to split up the deuterium. This gave us a particular temperature of the radiation at a particular time in the age of the Universe.

From this point on, Gamow assumed that the ratio of densities of photons to baryons would remain the same to this present day. As space expands, the photons and baryons become sparser and sparser. The densities of each reduce (but in the same proportion). The physics of black body radiation is well known and the photon density is related to the temperature. As the density of the radiation reduces, the temperature will fall and because of this, Gamow predicted that there should be a residual radiation left over from the Big Bang. This radiation is known as the Cosmic Microwave Background radiation, or CMB for short and he was also able to predict a value for the temperature of this radiation. There are several ways of doing this calculation but Gamow made an estimate of the present day baryon density (how much mass there is in each cubic metre of space today) and, using his assumption about the ratio of the densities of photons to baryons remaining constant, he could calculate the present day photon density and hence the temperature of the black body radiation. Gamow's team, in 1949, published the CMB temperature as being at least 5K and this prediction had risen to 7K by 1953. In his book published in 1961, Gamow predicted the CMB temperature as being 50K and so the scene was set for someone to come along and measure the actual value predicted not as 2.73K as is often cited, but at a predicted temperature of 50K.

We will return to the Big Bang theory later, as there have been many improvements to this theory since Gamow's time (for one thing, Gamow did not know about quarks). In any case Gamow's theory was seriously flawed, as high energy gamma rays in the very early universe would combine to form electrons and positrons, something not accounted for by Gamow. However, since Gamow was the first to make a 'good attempt' at how the universe was created, then he must be given the credit for the theory.

There was now a lot at stake as this was a theory that was uniting science and religion. A theory that told us how we all began, could tell us when it all began and how all matter formed. It was a theory that went beyond the laboratory walls and beyond the dry scientific journals. Everyone had an interest in how the Universe started and wanted to know more. Gamow's prediction of the CMB - the residual relic left over from the Big Bang, gave scientists something to look for. If they could find this low temperature microwave radiation pervading the whole of space then surely it would go a long way to verifying this strange seeming theory of how the Universe began. Or, looking at it another way, if they could show that there was no CMB then they could say for sure that it was wrong.

Chapter 10. Pigeon droppings and Nobel prizes.

Arno Penzias was born in Munich, Germany in 1933 to a Jewish family originating from Poland. His father worked in the leather business and this provided the family with a reasonable standard of living. However, this was the time of the Nazi party and so, in 1938, six-year-old Arno and the rest of the Penzias family, were rounded up, expelled from Germany, and put on a train to return to Poland. Unfortunately, or perhaps fortunately, if one considers the final fate of the Jews in Poland, the train was turned back at the Polish border as the deadline had passed and Poland had stopped accepting any more Jews.

On returning to Munich, Arno's father set about sorting out the papers in order that the family could immigrate to the USA. The British Government had humanely agreed to allow 10,000 Jewish children to go to Britain and Arno, along with his brother, were chosen to be a part of this exodus. His parents obtained their papers soon afterwards and joined the children in Britain, before immigrating to the USA, taking up residence in New York.

Arno Penzias studied Physics at University before spending two years in the army as a radar officer. He put this experience to good use and on leaving the army, gained his doctorate for building a Maser amplifier that he used in a radio astronomy project. The idea was to use the maser, which detects and amplifies microwave signals, to look for tell tale radio signals emitted by Hydrogen atoms in the space between galaxies. The microwave signals have a wavelength of 21cm and, if you find these, you have detected the existence of Hydrogen. Penzias pointed his maser here, there and everywhere and did not find anything at all, but he was still awarded his doctorate since it was not his fault - there just isn't much Hydrogen in intergalactic space to be detected. He went on to work at Bell labs where he met Bob Wilson, a Texan born in Houston, in 1936. The son of a Chemical engineer, Wilson had been an 'A Grade' student all his life, and for his doctorate, he worked in radio astronomy, mapping the distribution of Hydrogen in our galaxy.

Bell labs is a direct descendent from Alexander Graham Bell, the Scottish inventor of the telephone whose immortal words of "Mr Watson come here I want you" are embedded in the annals of Science. Not just because they were the first words ever to be sent electrically by Bell's new telephone, but also these words constituted the first emergency telephone call, since Bell had spilt acid all over himself and really did need the assistance of his lab technician, a Mr. Watson. Perhaps we should call Bell the 'patentee' of the telephone since there is some controversy over who actually invented the telephone. Several inventors were working on a telephone system and Antonio Meucci, an Italian who had immigrated to the USA, is said to have had a telephone system up and running five years before Bell patented his system - but the Italian could not find the money to patent the idea. Even so, Bell only just managed to patent the idea in the nick of time as Elisha Gray, another inventor of a telephone system, arrived at the patent office only a couple of hours after Bell had submitted his request for the patent. These must have been amongst the costliest two hours in the history of science.

Bell labs was then a part of the Bell Telephone Company and was the richest and biggest private research institute in the world, but it was allocated to the American Telephone and Telegraph company, on the break up of the Bell monopoly. Whilst the purpose of the labs was to find new technology that would benefit the parent company, there was also a strong tradition of pure research there too. The thinking behind this was that pure science would attract the best scientists and this would provide a 'spin off' for the telephone company.

It was at these labs that Shockley and others invented the transistor. There was an extremely sensitive antenna at the laboratories that was in the shape of a 'horn' for the detection of microwave signals and this had been used for satellite communications. At that time, communication satellites had been nothing more than a metal-coated balloon, which was inflated in orbit and reflected signals back to the ground. Consequently, the reflected signals returning to the ground were very weak and so a very large receiving horn was needed to detect them. As satellite technology had improved, such a large horn was no longer needed and, since the horn was still fairly new, it seemed a pity to throw it away - so a new research project was needed that could utilise the horn's capabilities.

The pair of researchers intended to use the antenna to detect radio signals from a burned out supernova, Cassiopea A, but there seemed to be a problem with the antenna. Wherever they pointed the horn, they detected an annoyingly low 'hiss' in the microwave region and this had to be removed before they could start their research. They thought that that it may have been caused by the equipment itself and checked and double-checked every joint and every connection until they were sure that the problem did not lie in the equipment. Next they checked objects such as the Sun and our galaxy itself but since the radio noise was the same in every direction no matter what time of day it was or what season of the year; it clearly could not be that. Being a large horn antenna, some pigeons had taken a liking to it and turned the inside of the horn into their home. The scientists decided that the warm pigeon droppings could give rise to a possible source of the noise and so they decided to remove the pigeons. Two weeks later, they finally managed to catch the pigeons, whereupon they were sent by company mail to a place far remote from the antenna and released. The horn was completely cleaned of droppings but unfortunately; the pigeons were of the homing variety and returned. The pigeons were caught once again, the horn cleaned once more but this time a gun was found and the pigeons 'sent to their maker' – the sacrifices one makes in the pursuit of science! But, to no avail as the radio noise still remained.

Meanwhile, just down the road at Princeton University, was a team of researchers, led by Bob Dicke. In the early forties, Dicke had built instruments that would detect microwaves and had been looking for background microwave emissions from distant galaxies. Their results showed that if this background microwave radiation did indeed exist then it had a temperature of less than 20K – otherwise they would have detected it. They published the results in 1946 and then moved on to other things. Nearly twenty years later, Dicke returned to this problem. He was a believer in a 'pulsating universe', one which 'bangs' and expands until gravity arrests the motion and it collapses back to a primeval fireball where everything goes back to square one before the bangs and expansions recur. He had asked one of his team, P.J.E. Peebles to find how the temperature of such a universe would vary as it pulsated and he came up with a present day value of around 10K for the CMB. Dicke and his team set about trying to measure this radiation and they were not alone, as several Russian teams had also entered the race to find the CMB. Dicke's team set up an antenna on the Physics department roof and, as they were doing so, Dicke received a telephone call from Penzias and Wilson saying something along the lines of "Could you help us with a little problem? We have this cosmic background radiation that won't go away!" Dicke knew what it was straight away and realised that he had been pipped at the post. The CMB had been discovered and all that was left for Dicke to do was to explain to Penzias and Wilson what they had found.

Since then, the temperature of the CMB has been measured time and time again and a special satellite COBE was sent up into orbit to detect ripples in the CMB. One of the problems of an expanding universe, is that one has to have irregularities where 'collapsing' takes place in order to form the clumps of material that form the stars and planets. If the original contents of the universe were evenly distributed, then this would not happen, everything would continue to fly further and further apart. However, if there were small variations in the early universe then areas of higher density would be the 'seeds' where galaxies and stars could form. When we look at the CMB, we are looking back in time, right back to the earliest times in the Big Bang and COBE is alleged to have found these seeds - which appear as slight temperature variations (around one part in one hundred thousandth) in the overall CMB temperature. For their work on discovering the Cosmic Background Radiation, Penzias and Wilson were awarded the Nobel Prize - well, a quarter of the prize each if one wants to be pedantic.

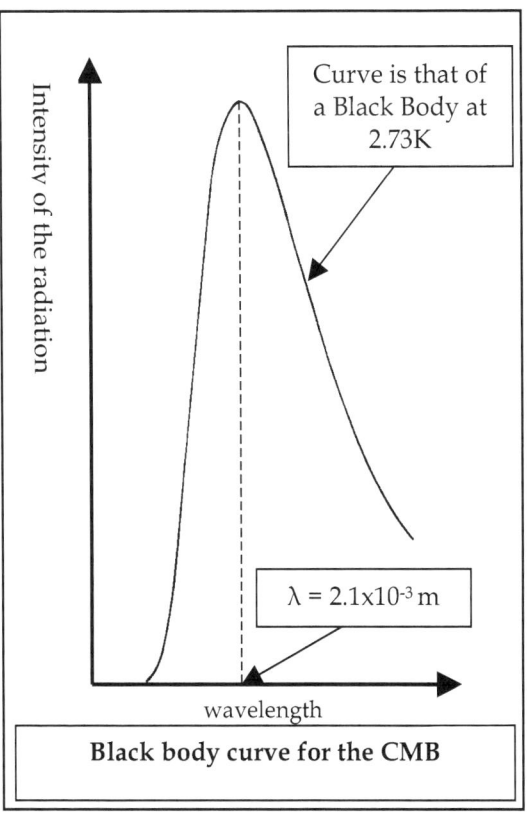

Black body curve for the CMB

The Nobel Prizes are named after Alfred Nobel, the Swedish discoverer of dynamite, who made a fortune out of the explosives and munitions industry. When he died, he stated in his will that his entire estate should be invested safely and the interest should be given to individuals who have made a significant contribution to the benefit of mankind. Whilst some of his relatives were not too happy about him having given his entire fortune away in this manner, we must not feel too sorry them because Nobel was a bit of a loner and besides, his brothers were oil magnates who, in any case, did not need the money. He designated five areas where these prizes should be awarded; Physics, Chemistry, Medicine, Peace and Literature – quite rightly deeming Mathematics as being unworthy of a prize! The three science prizes follow on from Nobel's scientific background, and the literature prize may well have had something to do with the fact that he had been a life long fanatic of the works of Shelley and had written several plays and poems himself. Cynics say that the peace prize was included as a sop to his guilty feelings, from the consequences of his discovery of dynamite, though any deaths from dynamite have mainly been through industrial accidents and not through acts of war. In 1968, an additional prize for economics was added to the other five, to be paid for by the Swedish Central Bank and awarded in Nobel's honour.

However, it is one thing to leave the money and quite another to decide who is worthy of receiving it. This was soon sorted out and the decisions are now made in strict secrecy (the details are only released fifty years after the prizes are awarded – no doubt to avoid any unpleasantness), presided over by the Royal family of Sweden and presented on December the tenth of each year – the anniversary of Nobel's death. Apart from the fame and respect given to a Nobel Laureate (the public can understand a Nobel Prize, even if they cannot understand the scientific achievements of the winners) the prize itself is a very substantial sum of money; each prize presently being over one million dollars and therefore not to be sneezed at. At least this is what Einstein's first wife Mileva thought. Einstein was on a scientific roll with his explanation of the photoelectric effect and special relativity so when they divorced in 1918, she had a clause put in the financial settlement of their divorce saying that if Einstein should be awarded a Nobel Prize in the future, then the money would go to her to bring up the children. A reasonable enough request until one reads the claims that Albert and Mileva are said to have had an illegitimate daughter, keeping the knowledge of this child a secret from everyone until thirty years after Einstein's death! However, when he received the Nobel Prize in 1921 this is indeed what happened and the money went to Mileva and the kids (would the rest of the children have starved had he not won the prize we ask? - so much for Einstein's theory of relativity and his relatives).

Whilst most agree that those who have received Nobel Prizes are worthy winners, it is often difficult in science to decide exactly who discovered what, as one discovery is built upon another, which is built upon another and so on until we achieve the final discovery. There have been cases of scientists feeling slighted by the committee - such as in the prize awarded for the development of MRI scanners, which are the ones used in hospitals to detect diseases. The prize was split between two worthy scientists, one American the other British, but a third scientist, also an American, felt that he had also been one of the 'discoverers' and took out a series of advertisements in National newspapers asking the Nobel awards committee to reconsider. They didn't.

There have also been one or two awards that in retrospect, might appear to have been misplaced. One prize, in 1949, was awarded to the surgeon who pioneered the removal of the front part of the brain in order to cure schizophrenia; a practice now deemed to be too severe. Another prize, in 1926, was awarded to a doctor who was said to have 'found' the cause for cancer and yet another, in 1927, for discovering that fevers cured mental illnesses. For this reason, prizes are now awarded well after the discoveries have been made to see if the discoveries withstand the test of time. This also leads to problems, as prizes cannot be awarded posthumously, that is, the recipient must be alive in order to receive it (the money is not much use to a dead person!). The Nobel Prize for DNA was awarded to Crick, Watson and Wilkins, but Rosalind Franklin, who did nearly all of the X-ray crystallography work, died of cancer before the prizes were awarded, and so 'missed out' on the glory.

The events leading up to the Big Bang theory were worthy of a Nobel Prize, but who to give it too? Whilst everyone seemed to expect Hubble to be awarded with a Nobel Prize, he died in 1953 and thereby foolishly disqualified himself by dying too soon. Einstein already had one and allegedly practical discoveries are given precedence over theoretical ones, so that ruled Gamow out of the equation (he died in 1968 anyway). This didn't leave the committee with many options and so it was that in 1978, the Nobel Prize in Physics was awarded to Penzias and Wilson.

Pyotr Leonidovitch Kapitsa was awarded one half of the prize for his "basic inventions and discoveries in the areas of low-temperature Physics" (nothing to do with the Big Bang). The other half of the prize was given to Penzias and Wilson "for their discovery of the cosmic microwave background radiation"; in other words, a quarter of a Nobel Prize each for their discovery.

This may seem a little strange, to be awarded such a prestigious prize for making a discovery that they were not looking for and did not know what they had found, until someone told them what it was, but the expanding universe was worth a prize and someone had to have it, if only a quarter of it.

Of course, at the same time, the CMB had also been discovered by millions of television viewers who were also not looking for it and nor did they understand what they had found when they had found it. This was in the days before twenty four hour television, when programmes were shown once over a period of eight hours instead of being repeated three times over a period of twenty four hours! Near midnight, the programmes would end and most viewers would quickly stand up from their seats and rush over to their television in order to switch it off, before the nightly 'Epilogue' appeared. This was a short message of hope and guidance on how to live one's life and was delivered to viewers by a solemn but religious person. Those who didn't switch the television off could stay up to discover the CMB. After the Epilogue, when the transmitters closed for the night (or even with present day televisions when they are 'off tune') random white dots appear on the screen and are known technically as 'snow'. These dots are caused by radio 'noise'; stray microwave background signals and about twenty percent of these are the CMB.

However, gaining international fame and fortune for discovering what television viewers had discovered long before is not particularly surprising. What is surprising, is gaining international fame and fortune for discovering something that had already been discovered! The CMB as 'blackbody' radiation from a source at 2.7K had been discovered and published long before Penzias and Wilson found it and long before Gamow had predicted the CMB temperature. It had been discovered in 1941 to be precise.

Andrew McKellar was a Canadian astrophysicist, and molecular spectroscopist who was born in Vancouver in 1910 but died in 1960. Since the late nineteen thirties, astronomers had noticed a series of absorption lines appearing at the blue end of the spectrum of light from distant stars. They reasoned that these were due to the absorption of starlight by molecules consisting of just two atoms and existing in the space between the stars. These molecules had to be within the cooler clouds occupying the space between the stars, since it does not take much energy to break up the molecules. There is therefore no way that they could exist in the hotter environment surrounding stars. One of these diatomic molecules was the cyanogen radical, (CN) and McKellar looked at which of the energy levels of this molecule could be causing the absorption lines. He realised that the molecule was already in an excited rotational state, before the absorption of the starlight, in the blue part of the spectrum took place. That is, the molecule had already absorbed a photon of radiation that had set it rotating before it absorbed a photon from the starlight. He reasoned that the radiation causing the original rotational excitation of the cyanogen molecules was blackbody and had a temperature of 2.3K. To have the molecule excited twice is a very large coincidence, unless that first excitation was from a radiation that existed everywhere. He reasoned that this blackbody radiation was the 'temperature of space' and in doing so 'pre discovered' the CMB at 2.3K twenty-two years before Penzias and Wilson executed the pigeons. However, since McKellar died three years before Penzias and Wilson's 'discovery' he could hardly raise his hand and say "I knew that twenty two years ago!"

McKellar's work was not widely known at the time and the Second World War would have hindered the knowledge of his results spreading throughout the scientific community but it was published, there for all to see and there is evidence that some of the big players in the Big Bang theory knew about it. However, we must not be too hard on Penzias and Wilson; they did carry out a very systematic search of the universe and kept meticulous records. Theirs was also the first direct measurement of the CMB, as McKellar had only found it by indirect methods.

Nonetheless, McKellar still discovered it first and we do not hear too much about it. The Nobel Prize was always out of the question since he died long before it was awarded, but why could we not add his name to the Big Bang hall of fame now? Remember the Hertzspung Russell diagram? Russell found it independently but Hertzsprung had already done the work and published it - so his name comes first. Why don't we do the same with McKellar? I don't know, but cynics put it down to the Big Bangers not wanting to do so. One of their biggest advertising campaigns about the Big Bang theory is based upon the belief that the Big Bang theory had predicted a CMB temperature of 2.7K before it was found by experiment. We already know that this was not the case, as Gamow's predicted temperature had risen to 50K by the time Penzias and Wilson finally 'discovered' it. If Big Bangers accept McKellar's original discovery then they would have to abandon the thought that the Big Bang theory predicted a result that the experiment then went on to find, one of the tenets of what represents a good theory. They would rather deny McKellar a place in history than give up their beloved ideas.

The other reason why the discovery of the CMB kicked up such fervour was that it was claimed that it had 'proved' the Big Bang theory correct. The Big Bang theory had a chief public rival known as the 'steady State' Theory. The main proponent of this theory was Fred Hoyle, a British scientist and mathematician who was born in Bingley, Yorkshire. Fred missed most of his early schooling due to truancy - whilst his mother thought he was at school, the school thought he was sick at home and meanwhile, Fred was doing his own thing. Eventually, he returned to the straight and narrow and managed to catch up enough on his studies to enable him to win a scholarship to Cambridge, where he studied mathematics.

Just as Fred's research career was starting to blossom, World War two broke out and Fred went to the Admiralty to work on radar. This was to prove extremely fortuitous as it was there that he worked with Hermann Bondi and Thomas Gold. The three of them discussed astronomy in their spare time and it was these three who, after the war was over, went on to propose the Steady State theory.

Fred was a great populariser of science, writing around forty books, a dozen of these being science fiction books. He also wrote a television serial and a children's play, both based on science. He put forward theories on Stonehenge being built to predict eclipses and that both life and viruses arrived from outer space. He also made five radio broadcasts on science, which proved to be so popular that they were repeated many times and also broadcast in the USA. In the last of these broadcasts, he poured scorn on the idea of the universe starting in a 'large fireball' and to make it look even more ludicrous, he referred to it as a "Big Bang". The phrase has stuck and so to recap; the Big Bang theory was born on April Fool's day, in a paper, with an author included in the list of names as a joke and with a name, 'Big Bang' christened to pour scorn on the whole idea. Big Bangers have even had conferences to try to come up with a different name for the theory but the only change they have been able to think of is the insertion of the word 'hot' at the front to make it the 'Hot Big Bang' theory. Hoyle would go on to explain the origin of the heavy nuclei in the universe, an explanation that is now generally accepted and this again was done in an attempt to disprove the Big Bang theory. Hoyle died in 2001, and never accepted the Big Bang theory. He, along with others, revamped the Steady State theory to overcome the original objections and was co-author of a heretical book, published in 2000, only one year before his death, outlining the failings of the Big Bang theory.

In the Steady State theory, as the universe expands new matter is created in the spaces between the old matter, with the result that the universe, as a whole, remains the same. In this theory, the temperature of space was predicted to be zero. The Big Bang theory, where the universe expanded from an original hot fireball required there to be a temperature of space greater than zero. Clearly, here was a way to distinguish between the two theories and to see which one was 'correct'. The discovery of the CMB meant that the 'Steady State' was wrong and *perhaps* the

Big Bang was correct. The discovery of the CMB did nothing to prove that other theories, such as Tired Light were wrong. It is a misnomer to say that finding the CMB proved the Big Bang theory correct; it didn't, as all it did do was to show that the old Steady State theory was wrong. Furthermore, the Steady State theory made a come back with a 'new' or 'quasi' Steady State theory, where iron fibres ejected from supernovae absorb the starlight, warm up and re-emit this heat as black body radiation in the microwave region. Indeed, a simple calculation shows that this results in a CMB temperature of 2.78K, much closer to the measured value than the 'predictions' of the Big Bang theory. However, it was too late. Minds had already been made up and school science curricula to this day, includes the teaching of how the discovery of the CMB 'proves' the Big Bang theory, whilst this is clearly not the case!

We will leave the CMB at this, point and return to the Hubble diagram and the role of supernovae as 'standard candles'. It was known that at greater and greater distances and higher and higher redshifts (with z above about 0.2), the Hubble diagram of redshift against velocity deviates from a straight line and takes an upward turn. To see what happens beyond this region needed us to find some way to extend the diagram to include the redshifts of bodies much further away. There is a limit to how far one can go with Cepheid variables as their apparent brightness becomes too small to be 'seen' at large distances. What we want is something that is really bright - and what can be better than supernovae, the explosion of a star!

Chapter 11. Bring on the inflating crystal spheres!

With its many successful predictions and the discovery of the CMB, we were told that the Big Bang theory had been proven beyond doubt - but behind the scenes heads were still being scratched. The problem was that it was all just a little bit too improbable for it all to have happened in this way and for life as we know it to exist. A lot of this was to do with how much mass there is in the Universe or, to be precise, what the density of the Universe is. As the Universe expands from an initial 'Bang' it will continue to coast outwards, but it will be slowed by the forces of gravity retarding the motion. If the amount of matter in the Universe is too small then the forces of gravity will not be strong enough to arrest the motion. The Universe will continue to expand at a slower and slower rate - but it will never come to rest. It will keep on expanding forever with the unpleasant thought that as all the matter moves further and further apart, the density of the Universe will, to all intents and purposes, reduce to zero! This situation is known in cosmology as 'open' curvature where General Relativity tells us that 'space' is curved outwards like the shape of a horse saddle. On the other hand, if the density of the Universe is too high then the forces of gravity will be strong enough to bring the expansion of the Universe to a halt and then cause it to collapse until it ends up in one great big Crunch! This situation is known as 'closed' curvature where 'space' is curved inwards like a football. The density which separates these two cases is called the 'critical density'. Should the density of the Universe be equal to this 'critical density' then the Universe would expand, gravity would bring it all to a stop and it would stay there, resting in this static but highly unstable state. Here space is known as 'flat' and we picture it as being like the surface of a pool table. One of the problems is that the Big Bang theory required the density of the universe be very close to this 'critical density' in order for life as we know it to exist – that is space should be 'flat'.

Should the density of the Universe be significantly less than the critical density then the gravitational forces would be so weak that galaxies, stars, planets and good old Earth would have been unable to form. A universe in which everything is whizzing outwards leaves no place for matter to come together and form these objects so desirable to our existence. No, if the density of the Universe had been too small then gravitational forces would just not have been strong enough to overcome the expansion and bring clumps of matter together to form the stars and planets we love so much. On the other hand, if the density of the Universe had been significantly greater than the critical density then the gravitational forces would have been so strong that they would have quickly arrested the expansion, caused the universe to collapse and brought the universe back to a resounding 'thud' - giving no time for the heavenly bodies and life itself to form. But what do we mean by the term 'significant'? That is, just how close to the critical density must the density of the Universe be, for us to be where we are, stood on a solid object caused by gravitational attraction, and yet given enough time for the Earth and so on to form (plus the time needed for Darwin's theory of evolution to get us out of the sea, into apes and then into humans) without gravity collapsing it all? It turns out that if the density of the Universe was more than 0.000000000000000000000000000000000000000000000000000000 00000001% away from the critical density then life as we know it wouldn't be here! Scientists argue about the odd zero but the end result is basically the same. It would be foolish to argue that this situation is so improbable that it could never happen since it really must have happened as we are actually stood here and doing our own thing on a clump called 'Earth'. What this result did was to cast doubt on the Big Bang theory and this 'little problem' was called 'the flatness problem'.

Another 'little problem' was that when the Big Bang occurred there should have been all sorts of weird and 'exotic' particles along with those that are commonly known about by the general public. Whilst some of these should have decayed by now, lots of others (called 'relic' particles) should still be here - and in such large numbers that we should be able to detect them with ease. The 'little problem' is that we cannot find them.

Another 'little problem' with the theory was that the early universe had to be almost exactly the same in every place (homogeneous). This came from observations of the CMB which is all encompassing and basically identical in any direction one cares to look. When we look at the CMB, the expanding Universe theory tells us that we are looking back in time and seeing the early universe and the start of things to come. In order for galaxies and so on to form, there must have been regions of space slightly denser than others. As we saw before, if matter was evenly spread out then gravity could not cause some areas to coalesce into clumps to form the stars, planets and galaxies. That is, after the 'Bang!' there just had to be some very slight variation in the distribution of energy and hence 'matter' and then gravity would do the rest. It would bring these areas together- which in turn would attract even more matter until, from these early 'seeds' of slightly denser regions, the galaxies, stars and planets would eventually grow. We see these 'clumps' in the CMB as slight variations in temperature and it is thought that these are the seeds of the structure of the Universe – part of which we now call home. The clumps in the CMB vary by less than 0.0001% and, since we are looking back in time when we look at the CMB, the early Universe must not have varied by more than this amount i.e. it must have been very uniform indeed. This leads us to another little problem because in a 'Big Bang' we are saying that the whole of the bang and expansion must have started everywhere at exactly the same moment in time – which is also pretty unlikely.

Along these same lines lies another 'little problem' – well it is a 'great big problem' really and is known as the "Horizon Problem", which tells us that if the Big Bang theory is correct then at some time or other, radiation must have travelled faster than light! The problem is that the CMB is at the same temperature everywhere in our Universe and it is thus in 'thermal equilibrium'. Places that were originally 'hot' and those that were originally 'cold' have exchanged radiation so that they are now at the same temperature. If our Universe started in one point and expanded outwards the 'diameter' now, or to use Big Bang terminology, the distance from one horizon to the other will be about 28 billion light years. But the Universe is only a little under 14 billion years old and so the furthest that any radiation could have travelled in this time is 14 billion light year - so there is no way that any radiation can have travelled from one edge of the Universe to the other to keep them at the same temperature – unless it travelled at a speed greater than that of light!

Despite all these 'little problems' we were still all being told that the Big Bang theory was correct with no proviso attached to say that perhaps everything wasn't quite as hunky dory as we were led to believe. What was needed was a 'tweak' to the theory and that 'tweak' came in the form of inflation.

The idea of 'inflation was worked out separately by Alexei Starobinsky in Russia and Alan Guth in the USA - though it was Guth who gave us the term 'inflation' in 1980. General relativity tells us that the rate at which the Universe expands increases as the 'energy' density increases. The 'energy density' is how much energy there is in total (due to matter, radiation and fields) per cubic metre of space. In electric or magnetic fields, forces act which make things accelerate and gain energy and so there must be energy associated with the field. In the Big Bang theory, we do not know what happened at the very beginning because our present physical laws do not apply, but in the Inflation theory, it is assumed that there was a type of field called a 'scalar field'. A scalar field has strange properties in that when the field is 'strong' and the energy is highly concentrated, a scalar field produces rapid, exponential expansion of space. But remarkably, whilst space expands rapidly, the energy density of a scalar field hardly changes at all. With matter, if the volume occupied by that matter doubled then the density would halve. But not so with scalar fields, as here the energy density hardly reduces from its former value – until we reach a certain 'cut off' value. When the energy density has finally reduced to a certain value, then it falls off very rapidly to give us the very sedate expansion rates that we see today. It must be said that we have never detected such a field but theorists believe them to exist so who are we to argue? In any case, we do not know what was happening at the very beginning so anything is possible (or so Big Bang codsmology tells us).

Apart from not knowing where the scalar fields come from, or not being able to detect any scalar fields other than in a theoretician's head, they do solve the 'little problems' with the Big Bang theory (sceptics say that they just replace one problem with another!). Firstly, since we see very little variation in the CMB measurements, the early Universe must have been very uniform indeed. This meant that in the old Big Bang theory, expansion must have started everywhere at exactly the same time. Inflation gets over this by saying that we now only need one tiny region where the energy density of the scalar field was high enough to start the rapid exponential expansion and this region would take over. That is, this tiny region where the exponential expansion began would become larger and larger to such an extent that it would very quickly become the whole of the universe as we know it. There would be some fluctuations in the density (which would eventually become the structure of the Universe) but the rapid expansion caused by inflation would ensure that these fluctuations became bigger and bigger so that they would form the 'large scale' fluctuations in the CMB. Inflation would also reduce the overall variation in these fluctuations so that they hardly vary at all. Imagine the carpet in the hallway of your house. When the carpet becomes rumpled we pull one end and the rumples stretch lengthwise and become hardly noticeable. It is the same with the fluctuations in the Universe – stretch them and they become less noticeable i.e. homogeneous.

Secondly, inflation explains why the density of the Universe is so close to the critical density – that is, inflation solves the little problem of "flatness". Whichever way the initial Universe was curved, the rapid stretching of space meant that it would be stretched into a 'flat' universe. The Earth's horizon looks 'flat' because the radius of the Earth is large and it is the same with the Universe. As the Universe expands rapidly any curvature is stretched so much that it becomes 'flat'. Again, think of the hall carpet, as we pull one end and stretch it, any curves in the rumples in the carpet smooth out. They become less 'curved' and we end up eventually with a very 'flat' carpet.

Thirdly, inflation explains why we cannot find the 'relic' particles – those weird particles that were left over from the Big Bang that should exist everywhere in their billions. Since it was only a very tiny region of the universe that experienced this rapid expansion and eventually became the whole of the Universe, then we only expect to find those particles that where originally within this small region of space. This tiny region only contained a few of the relic particles and so, since this region has now expanded to form the whole of the Universe, the density of the relic particles in the Universe has now become almost zero - with the result that we do not expect to find any! We can use our analogy of the hall carpet once more. What inflation has actually done is to sweep the relic particles under the carpet so we don't expect to see them!

Fourthly, the Inflation theory could explain why the fluctuations in the CMB occurred in the first place. It is all to do with quantum mechanics and this predicts that there will be small scale fluctuations in the energy density. As inflation causes expansion these fluctuations increase in size. Quantum mechanical fluctuations occurring at the start of the inflation period will be stretched so that they are now huge and occupy most of space. Quantum mechanical fluctuations that occurred near the end of the inflationary period will not be stretched as much and thus they still appear, but as the 'small scale' fluctuations in the CMB. The Inflation Theory could predict how these fluctuations were distributed and, towards the end of the twentieth century, these fluctuations were said to be discovered by the Wilkinson Microwave Anisotropy Probe (WMAP) - a satellite probe sent up to map out the CMB. With this successful prediction, inflation was welcomed into the Big Bang codsmology. However, as we shall see later, in 2005 it was shown that the larger of the fluctuations in the CMB were aligned with our own solar system - which was just not in the inflationary plan at all!

And lastly, inflation solves the 'horizon problem' since it was only a tiny 'bubble' that inflated to become the whole of the Universe then we expect it to be all at the same temperature and in thermal equilibrium. The only problem is that nobody knows why the tiny 'bubble' inflated in the first place.

Big Bangers now believe the beginnings of the Universe as being, in principle, something along these lines:

What happened at the very beginning of the Universe, that is in the 'Big Bang' itself, is still open to speculation and usually involves 'new' or 'unknown' Physics and so we will start our story about one ten-million, million, million, million million, million, millionth (10^{-43}) of a second after the Big Bang. The Universe is very hot and dense at this point with a temperature estimated to be about 10^{32} K (which is HOT!) and what is now the entire Universe, existing as energy in no describable form, is contained within a tiny 'speck'. Or, to put it another way, hundreds of billions of galaxies squashed into a space smaller than the nucleus of an atom. Inflation then took over and the Universe expanded at an unbelievable rate - what do we mean by an 'unbelievable rate? Well it expanded by a factor of 10^{50}, so if you imagine a speck 10^{-32} cm across (one million, million, million, millionth the size of an atom, then it grew to the size of the whole Solar system in 10^{-33} of a second! Yes, that was faster than the speed of light! Inflation ended at this point and from this energetic gas, photons, quarks, antiquarks, gluons and electrons formed. The density of the Universe is around 10^{25} kg/m^3, which is hard to appreciate until you realise that it is like having 200,000,000 of the good ship 'Titanic' all squashed into a cube with sides 1mm. Or, to put it another way, the Earth would be the same size as your television set!

What we have is a great ball of 'fire' that starts as a small point in space or 'an infinitesimal singularity' and from then on expands outwards all the time. This great 'fireball' represents all there is. It is both the Universe and the whole of space itself, expanding into nothing because nothing but this fireball exists. As the fireball expands, it is space itself that is being stretched, carrying the particles along with it. Whilst this expansion has no real effect on the actual particles (they don't increase in size, they just move further apart), it does affect the photons of radiation that, at this point, belong to the gamma ray part of the electromagnetic spectrum due to their extremely high energies.

As space expands, the photons are stretched and their wavelength increases and hence they are 'redshifted'. This requires their frequency to become less. Lower frequency means less energy and so the 'primeval fireball' that is our Universe becomes 'cooler'.

From around 10^{-35} seconds until 10^{-12} seconds, we had the 'quark-gluon plasma' but gradually, as the Universe cooled, quarks could come together in their threes to form the protons and neutrons in the Universe. Protons and neutrons are each made up of combinations of three quarks but until this time the quarks had far too much energy to be confined into such a small space as the nucleus of an atom and so the quarks existed as individual particles. Nowadays we do not find individual quarks, as the conditions have to be extreme for this to happen – the temperature here is nearly 200,000 times hotter than the core of the Sun. However, eventually, around 40 microseconds after the Big bang, this quark-gluon plasma cooled sufficiently that quarks could come together in their threes to form the protons, neutrons and their anti particles. These particles are whizzing about and bashing into each other at every opportunity. In doing so, they are continually changing from one form to another by a process known as 'pair production'. If two photons come together and interact, then they can form a particle and its antiparticle. In the same way, if a particle bashes into its anti particle then they will annihilate each other to produce 'pure' energy in the form of two photons.

There came a point (at about 10^{-7} seconds) where the expansion of the Universe had stretched the photons so much that once created, the photons were now unable to reform the original particle and antiparticle that had brought them into existence in the first place. Before, a proton and an anti proton or a neutron and an anti neutron could collide, annihilate each other and form two photons and then the two photons could collide with each other and reform the particle and its anti particle - but with the expanding of the universe and the stretching of space, all this had to come to a stop because the photons were becoming less and less energetic. After this point, when a particle and its anti particle meet and annihilate each other to produce two photons, the photons stretch a little more and they do not have enough energy left to recombine and reform the original particle and its anti particle.

Fortunately, there were slightly more particles than antiparticles and so they did not completely annihilate - thus leaving us with some particles with which to build the Universe. The Universe has now cooled to a mere 10^{13} K and the density has fallen to 10^{17} kg/m^3. This means that the density is like having our 200,000,000 'Titanics' that used to occupy a cube of side one millimetre, now squeezed inside a suitcase.

When the Universe was around three seconds old, it came the turn of the electrons and their antiparticle (the positron) to annihilate each other and once again there were slightly more electrons than positrons so we ended up with a least some electrons to form the atoms of our Universe. By three minutes nucleosynthesis had begun. The temperature had become 10^9 K and the Universe now the same density as that of the Titanic in 1912 - just before it sank. Deuterium and Helium could be formed as set out by Gamow but we had to wait until the Universe was a mere 300,000 years old before these heavy nuclei could pick up and retain electrons to form neutral atoms. The Universe is now one thousandth of its present size but photons still had enough energy to rip off any unsuspecting electron that dares to attach itself to a potential atomic nucleus.

As time goes on, the Universe expands and expands and the photons are stretched further and further. Since they are more and more redshifted, their energies become less and less and we reach a point where the photons do not have enough energy left to dislocate an electron from an atom. The Universe is now at the grand old age of one million years old and the temperature has fallen to a cool 4000K. The density of the Universe has fallen so much that where it was once equal to the density 200,000,000 Titanics squashed into a space of one cubic millimetre it is now equivalent to the density of a single ship stretching one third of the distance from the Earth to Venus.

This is a nice part of the story because this is where neutral atoms drop out of the 'melting pot' and our familiar atoms are formed. Photons do not tend to interact with neutral atoms and thus our great fireball ceases to exist as such. No longer is our Universe like a giant 'yellow star'. It is now, transparent. The photons no longer interact with matter and this is known as 'the decoupling of radiation from matter' or simply, 'decoupling'. This marks an important step in our road to the Big Bang, as this is the birth of the 'Cosmic Microwave Background' or CMB for short. Whatever photons are left at this stage of the Universe remain as photons since they do not have enough energy to combine and form new particles. Granted other radiation will be formed in stars and supernova in the future that can interact, but this is not part of our CMB. This radiation will be extra. The Universe continues to expand at its relentless rate and space, along with the photons inside, it will stretch further and further. The photons forming the CMB will become less and less energetic and thus the Universe will continue to cool. As for the rest of the Universe, the distribution of matter is not entirely 'even' or 'smooth' as it has 'fluctuations' in it. That means that some bits are denser than other bits. Gravitational forces cause the denser parts to start to 'clump together to form gas clouds, stars, galaxies and all the other bits and pieces that make up the Heavens in our familiar night sky. The Universe continues to expand carrying these galaxies along with it, so that they give the appearance of speeding away from each other with the redshifts and 'velocities' as measured by Slipher and Hubble. They are not actually moving relative to space, it is space itself that is 'stretching' carrying the galaxies along with it. The prehistoric Photons that make up the CMB, that is, photons left over from the original fireball, continue to be stretched and stretched, becoming 'cooler and cooler' as time goes on.

The stars and galaxies formed by the clumping of matter are known as first generation stars since they only contain the lighter elements. They contain no carbon or iron; these are elements that are formed later. Once these first generation stars have come to the end of their lives they will die. During their death throes, they will burn heavier and heavier atoms in their cores and, if the mass is great enough, they will form the heavier atoms by nuclear fusion. When the star explodes in a supernova, these heavier elements are dispersed throughout the universe as dust, which can then form new, second-generation stars. Even now, younger stars do not contain any iron, whilst our Sun does contain iron. This tells us that our Sun is a second-generation star, made up of second hand parts from supernova remnants. We must not feel too upset about this, since we too had the same beginning. Life is carbon based and the carbon was produced in supernova explosions. What was it somebody said about us all being stardust?

Chapter 12. Now, this is what I call a Big Bang!

When we looked at Goodricke and the Cepheid variables, we saw that a star in the prime of its life produces energy by a process involving the fusing together of protons to produce Helium atoms plus some energy (A Helium nucleus consists of two protons and two neutrons). The energy produced takes the form of heat which gives rise to 'radiation pressure' acting outwards.

In the prime of a star's life, the outward radiation pressure is counterbalanced by inwardly acting gravitational forces, with the consequence that the size of the star remains constant.

As the star ages, its supply of Hydrogen begins to dry up and so the star finds itself in a situation where the forces no longer balance. What happens next, is that the star expands and cools and, since it becomes large and has a red colour, it is imaginatively called a 'red giant'. The star changes its fuel supply and whilst Hydrogen is still being burned to produce Helium, the heavier Helium congregates at the centre. When there is enough Helium collected at the centre, the fusion of Helium nuclei 'kicks in' and the energy produced blows away the outer layers of Hydrogen.

One might have thought that the star's next step would have been the fusing together of two Helium atoms, but not so. This is an old star with lots of experience in nuclear fusion and so it decides to be adventurous and goes for 'three at once'. That is, the next step is to fuse three Helium atoms together at the same time to produce the element Carbon (there is no stable atom with eight nucleons anyway). The nucleus of Carbon has six protons and six neutrons. It might appear that to make these three Helium atoms do a 'three in a bed' is a pretty unlikely scenario, and it would be if it were not because of a special property of the Carbon nucleus. The Carbon nucleus has an energy level that is filled perfectly when three Helium nuclei come together.

Fred Hoyle and Edwin Salpeter were the first to realise that stars go straight from the lighter Hydrogen and Helium to the heavier Carbon. This also leads to the creation of Oxygen within the red giant, when one Carbon nucleus combines with one Helium nucleus to give the eight protons and eight neutrons that make up an Oxygen nucleus. It is by this process that both the Carbon from which we are all made and the Oxygen that we breathe is made. But this is where the star faces a problem. As more and more of the smaller nuclei fuse together, there are fewer and fewer nuclei left within the star. The chances of these larger nuclei bumping into each other and fusing to create even bigger nuclei becomes less and less. To increase the chances of nuclei bumping into each other later on in life, a star needs more nuclei to begin with. That is, what happens next depends upon the original mass of the star. For small stars, like our Sun, the fusion goes no further than the creation of Carbon and Oxygen. It throws away its outer layers as a cloud of gas whilst the inner core collapses slowly under the inward pull of gravity.

Having cooled, the radiation pressure is no longer able to compete with the relentless forces of gravity. It does not collapse forever though, as we reach a point, when the size of the star is about the same size as the Earth, where quantum mechanical effects produce forces between the electrons and these forces counter the forces of gravity - allowing the star to remain at this Earthly size. What follows is a 'boring' death whereby the star, now known as a 'white dwarf' just gradually cools and fades away. We will not be here to witness the expiry of our Sun though since we on Earth will have been frazzled away during the red giant phase!

Big stars having an original mass larger than that of our Sun have a different fate. Once these big stars burn out and collapse no quantum mechanical forces between the electrons can save them. The gravitational forces are just too big. The star collapses; carbon and oxygen continue to fuse until we end up with iron; this is then the end of the road for fusion. The elements form layers and the star begins to resemble an onion (oh no, not those Greeks again!). The central core is iron, then silicon, and then oxygen with an outer skin of carbon. What we end up with is a nuclear bomb of tremendous proportions with the energy released in fusion given out as a brilliant burst of energy known as a 'supernova'. The larger the mass of the original star, the bigger the explosion. As such, whilst they are very bright and can be seen from far off, they are of little use for distance measurement because they are not 'standard candles' - since they all have a different true brightness.

However, there is a 'critical mass' for a star that separates the two cases. Below this critical mass, the star will just fade away as a white dwarf but above this critical mass the star will explode as a supernova. This critical mass is known as the Chandrasekhar limit, named after Subrahmanyan Chandrasekhar who worked it out originally and this limit is 1.4 times the mass of our Sun. Chandrasekhar was born in Lahore, India, in 1910 and he is the nephew of C. V. Raman the famous Physicist and Nobel Prize winner. His uncle had studied in England and on a later visit he had been inspired by a lecture by Lord Rayleigh on the scattering of light and the resulting explanation of 'why the sky is blue'. On the boat trip home, via the Suez Canal, the many different colours produced by the Mediterranean Sea had fascinated his uncle. So, on his return to India, he set about investigating the effects of water scattering light and producing these colours. From these small beginnings, Raman went on to discover a type of scattering of light which now bears his name, 'Raman scatter' and for this he was awarded the Nobel Prize – and not a share in one, he received a whole one!

Inspired by his uncle, Chandrasekhar studied to be a theoretical astrophysicist at Madras University and was given the opportunity to complete his studies at Cambridge, England. Following in his Uncle's footsteps (or should we say, wake?) whilst on the boat from India to England, he applied his knowledge of special relativity to his interest in white dwarf stars. At that time, all stars were thought to end their lives as white dwarfs but Chandrasekhar realised that if the mass was above a certain limit, then the star would not just fizzle away but collapse in upon itself. Once at Cambridge, he completed his doctorate but did not forget his work on white dwarfs. He presented his findings at 'The Royal Astronomical Society' in London, where it was received with instant derision! Possibly as a result of this, he moved to the USA and joined the University of Chicago, where he remained for the rest of his working life. Up until Chandrasekhar's work, it had been thought that all stars just faded away into obscurity, but he showed that the larger stars came to a much more dramatic end, with the possibility that very large stars could collapse into Black Holes; objects so dense that even light cannot escape. Consequently, a star with a mass of more than 1.4 times the mass of our Sun will end its life suddenly and painfully as a supernova, whilst those with a mass of less than this, will end their lives in a slow but peaceful death as a white dwarf. Chandrasekhar was awarded the Nobel Prize (well half of one) "for his theoretical studies of the physical processes of importance to the structure and evolution of the stars" in 1984 and thus completed the Uncle and Nephew Nobel Prize double act. It is worth noting that the boat trip between England to India has been responsible for at least two Nobel prizes - perhaps we should all give up flying and take to a boat instead!

Unfortunately, it is not as simple as that. Our Sun leads a lonely life, as it is a single star. Many stars are a part of a star 'system' consisting of two or more stars interacting gravitationally with each other with the result that they waltz around each other. For combinations of stars, such as a binary pair, when the time comes for these stars to die, they will not necessarily both die at the same time. As in life, one partner is left to grieve once the other has gone. An example of this is Sirius, the Dog Star. Sirius is the brightest star in the night sky and its name is most likely derived from the Greek word meaning "sparkling". Sirius lies at the foot of its master, the constellation of Orion the hunter. Sirius is one of a binary pair but the second star has 'passed away' and has become a dim white dwarf, sometimes known as the 'flea of the Dog'. The distance between the two stars is such that there is no interchange of matter between them.

It is possible that if the stars were closer together, then the still alive and burning gassy star could lose some of its outer atmosphere to the white dwarf. This would mean that the mass of the white dwarf would increase and at the moment that its mass reaches the Chandrasekhar limit, it turns into a nuclear bomb and explodes as a supernova. In each such case, the supernova will burn with a predictable brightness as they all have the same mass, 1.4 times the mass of our Sun. We name them 'supernova Ia', and here we have a standard candle, or more truthfully, a 'standard nuclear bomb', which extends our distance ladder much further out into the universe. When supernova Ia's go off, they burn with a brightness equal to five thousand million of our Suns burning at the same time and they can be seen five hundred times further away than the brightest Cepheid variables. They extend the distance ladder to around three thousand million light years and bring us close to the 'frontier of the universe as we know it.

In order to calibrate a supernova Ia, we must find galaxies that contain both Cepheid variables, and the supernova. One such example is the cluster of galaxies in the constellation of Fornax in the Southern hemisphere. This cluster of galaxies is, to all intents and purposes, the same distance away and contains thirty-seven bright Cepheid variables. By measuring the time period of these Cepheid variables, we have already seen how we can determine their distance and hence the distance to the star cluster itself. This distance is found to be around sixty million light years. The Fornax cluster also contains two supernovae Ia and so if we measure their apparent brightness it will give us a means of calibrating the supernova standard nuclear bomb and thus enable us to determine the distance to other galaxies. By comparing the apparent brightness of supernovae in distant galaxies, to those in the Fornax cluster and using the inverse square law, we can find how far away the distant galaxy is.

Unfortunately it is not as easy as that. When one of these supernovae explodes, it starts off fairly dim and the brightness builds up to a maximum after about three weeks, when it will be brighter than the whole of the host galaxy. The brightness then starts to reduce slowly over the next year or so. Firstly, whilst these supernovae all have a similar maximum brightness they do differ. Secondly, it is not always possible to detect a supernova as soon as one occurs. It is often the case that when the supernovae are identified and, more importantly, when telescope time is available, the supernova is already past maximum brightness and dimming slowly. There is a large demand for the huge telescopes needed to measure light output from these very distant events and they are also in demand for other fields of research too.

Astronomers have to bid for telescope time long in advance, and one cannot say "Ooh look, a supernova, I need the telescope now", since the astronomers using the telescope that night might have waited many months for their turn and may be reluctant to give up their time allocation. What happens, is that when a supernova team is allocated telescope time, a second telescope having a wider field of view is used and, in the same way as Leavitt looked for Cepheid variables, pictures of distant galaxies are compared with previous ones of the same area to look for new bright 'events' that could possibly be a new supernova. Nowadays, computer programs do the searching with the final decision as to what is a supernova and what is 'noise', being made by an experienced pair of eyes. The supernovae team then observes the targeted supernova and the apparent brightness of the supernova is measured over a period of time. A graph of 'brightness' against 'time' is drawn, to give the 'light curve' for the supernova. Of course, not every supernova found is a supernova Ia but these can be identified - again by their light curves. At the time of writing, astrophysicists are finding around eleven supernovae Ia every month.

Over the years, astrophysicists have been able to calibrate these light curves so that even if they 'missed' observing the supernova at maximum brightness, they can work back from the light curve showing how the supernova dims with time to find out just how bright it was at maximum brightness. There is a second benefit from working from the light curves, rather than measuring the maximum brightness of the supernova. The maximum brightness of all supernova Ia are similar but they do differ - and differ too much for the precision we wish to achieve in the determination of the Hubble constant and hence the age of the Universe. Fortunately, the light emitted by dim supernova Ia is redder and decays more quickly than the light emitted by a bright supernova Ia, which is bluer and decays more slowly. This shows up in the light curves and again, by examining the nature of the light curve we can overcome the variation in output between supernovae and determine the brightness at maximum.

To sum up, in order to use supernovae Ia as distance indicators we:

i) Find a supernova to observe.
ii) Measure the apparent brightness over a period of time.
iii) Plot a graph of apparent brightness against time (a light curve)
iv) From the light curve, we can tell if it is a supernova Ia.
v) If it is, work back from the light curve to find the apparent brightness when the supernova Ia was at maximum brightness.
vi) Compare this maximum brightness to supernovae Ia in galaxies that contain Cepheid variables and hence at a known distance away.
vii) Use the inverse square law to calculate how far away the supernova is (and allow for the effects of dust etc.).

Thus we are able to find how far away the supernovae are. The redshift is also measured by measuring the shift in the absorption lines in the starlight from the host galaxy or by measuring the shift in emission lines emitted by the supernova itself. Thus we are able to extend the Hubble diagram far out into space and see if the redshift (or 'velocity' as Big Bangers like to interpret it) remains proportional to the distance over even greater distances than before.

Chapter 13. The farmyard of codsmology!

Relatively nearby Supernovae, astronomically speaking, have proved to be a precise way to measure both the redshift and distance to galaxies and we have been able to use them to determine an accurate value for the local Hubble constant. From the measured redshift, cosmologists have calculated the respective recession velocity and plotted the results on a graph of velocity versus distance. The resulting Hubble diagrams have become closer and closer to the perfect line that we have been told to expect. In 1996, published results of supernovae Ia data gave a Hubble diagram with such a spectacular straight line that it led some to state that it "provided dramatic confirmation of the Hubble law" – which it surely did when using 'nearby' supernovae. They plotted a graph of 'velocity' against distance and the slope or 'gradient' of this graph gives the Hubble constant, H. These results gave a value of the Hubble constant of 64+/- 3 km/s per Mpc.

The 'megaparsec' is an archaic unit of distance left over from the days when parallax was used in astronomy to measure distances to nearby stars. Why cosmologists still prefer to use this unit instead of the 'light year (ly) or even the standardised unit of distance, the metre (m) to measure distance is anyone's guess, but perhaps they like to use the megaparsec in order to make them feel 'different' and perhaps 'better' than anyone else. One megaparsec is one million parsec and the name 'parsec' is an abbreviation for 'parallax second'. Distances within the solar system are measured by astronomers in 'Astronomical Units' (AU), which is the average distance from the Earth to the Sun. Distances outside the solar system are measured in 'parsec' and this is defined as "the distance from which the average radius of the earth's orbit would subtend an angle of one second of arc". There are sixty minutes in a degree and sixty seconds in a minute and so an angle of one second of arc is one three hundred and sixtieth of a degree. Imagine that you are standing on an object one parsec away. When you look at our solar system, the angle between our Sun and the Earth will be one three hundred and sixtieth of a degree.

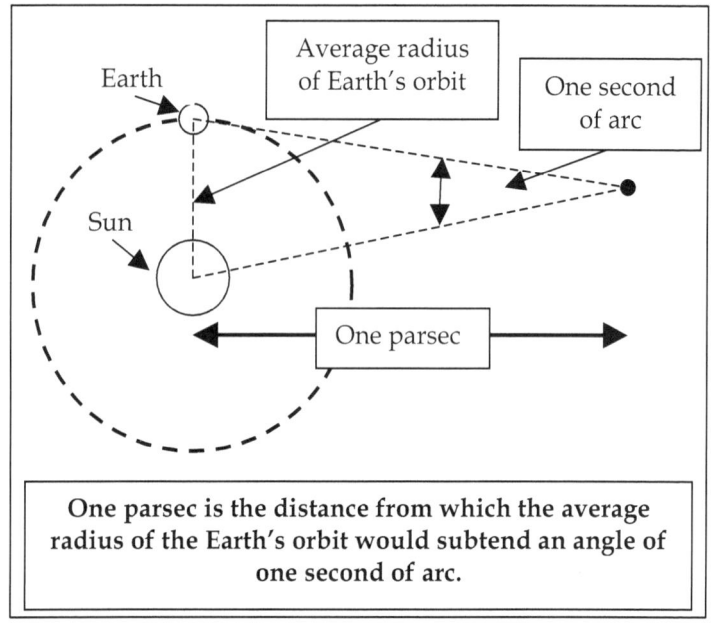

One parsec is the distance from which the average radius of the Earth's orbit would subtend an angle of one second of arc.

In an expanding Universe, the Hubble constant is supposed to tell us the rate at which the Universe is expanding. A Hubble constant of 64 km/s per Mpc means that an object one megaparsec from earth is moving away at a velocity of 64 km/s. An object two megaparsec from earth is moving away at a velocity of 128km/s and so on.

Distance to object in Mpc	Recession velocity in km/s
0	0
1	64
2	128
3	192
4	256

These are strange units, so why don't we, at least, be scientific and use the Standard International (SI) units - measuring quantities in metres, kilograms and seconds etc. To do this we must change the 'km/s' to 'm/s' and the Mpc to metres. One megaparsec is equivalent to 3.09×10^{22} metres, and so a Hubble constant of 64 km/s per Mpc is the same as an expansion rate of 2.07×10^{-18} 'm/s per metre'. The value is much less since the metre is much smaller than the megaparsec. This means that an object one metre from us will be moving away at a velocity of 2.07×10^{-18} m/s whilst an object two metre from us will be moving away at a velocity of 4.14×10^{-18} m/s and so on. Now, we know that it is really space itself that is stretching carrying matter along with it, so our result is telling us that a piece of space one metre long is stretching or increasing in length by 2.07×10^{-18} metre every second.

This precise result also gives us an excellent idea of the age of the Universe. The age of the Universe, or 'Hubble time', is the reciprocal of the Hubble constant.

> Age of Universe = 1/(Hubble constant)

Here, we must use the Standard International system of units and so instead of the Hubble constant being 64 km/s per Megaparsec, we take the value of H as 2.07×10^{-18} m/s per metre. Substituting this value gives an age of the Universe as fifteen billion years.

But, wait for just one moment. The units of H, 'm/s per metre', are very strange indeed. The unit, 'm/s per metre' is really 'metres per second divided by metres' – and so the metres cancel! The Hubble constant is not a velocity at all; it only has units of 'per second'.

> $H = 2.07 \times 10^{-18}$ 'per second'

And, wait for another moment. The magnitude of the Hubble constant is 2.07×10^{-18}. That cannot be right! It is just too much of a coincidence! This value, 2.07×10^{-18} per second is the same as 'hr/m for the electron in each cubic metre of space'. That is, what our results are telling us is that the Hubble constant is equal to 'this much of an electron' in each cubic metre of space and 'this much' of the electron is exactly 'hr/m'. These three constants, the Planck constant h, the mass of the electron, m and the radius of the electron, r, are very common and can be found stored in the memory of any schoolchild's scientific calculator. What this means is that scientists have spent the last eighty years looking for the value of a constant that any schoolchild could have found in seconds by pressing a few keys on their calculator!

Now, it must be said that the value of the Hubble constant of H = 64 km/s per Mpc cited above is just one of many values found by different workers using various techniques. Many scientists choose to use a value for H found by an international group of scientists who spent eight years using the Hubble space Telescope (HST). They came to the conclusion that the Hubble constant had a value of 72 +/- 8 km/s per Mpc. This means that the 'best guess' for the value of 'H' is 72 - but it could lie anywhere between 64 and 80 km/s per Mpc. Consequently, this value of the Hubble constant, found by an international team of scientists using the Hubble Space telescope over a period of many years, is still consistent with the value that our schoolchild could have pulled up on their calculator in a matter of seconds! In SI units this value of H is $(2.33 +/- 0.26) \times 10^{-18}$ s^{-1} compared to the value of hr/m per metre cubed of space of 2.07×10^{-18} s^{-1}. In order to simplify things perhaps it would be a good idea to name the quantity "hr/m per cubic metre of space" as a constant in its own right and give it its own symbol. Lets assign the symbol 'A' to this quantity, 'A' for Ashmore's constant (look, it's my theory, I found it and more importantly, it is my book so I will do as I like!).

Let A = hr/m per cubic metre of space.

The constant 'A' has units of 'per sec' or s^{-1}. Consequently, the value of the Hubble constant from the supernovae results is just 'A'. The HST result cited above is H = (1.1 +/- 0.1)A. In March 2006, a group of scientists led by Alan Sandage and also using results from the HST, published a value for H of 62.3 +/- 1.3 km/s per Mpc, or 0.97A. But which value is correct? The answer is that they are all correct; we often get different values when we measure things in different ways. One way around this is to find the average value of all the most recent results. When I wrote my original scientific paper, in order to get an unbiased sample, I fed the title words "Hubble AND constant AND measurement" into the ADS database search engine and chose 'return 100 items.' The 'ADS Database or to use its proper title 'NASA's Astrophysics Data System Bibliographic Service' is a fantastic service were individuals can access millions of scientific papers over the Internet and this database contains just about all of the scientific papers of note. Of the papers listed, I took all those giving an actual value for H and these are shown below. In theory, all the most recently measured values of H over the previous 5 years (at the time of writing the paper) should be listed. The values are given in terms of 'Ashmore's Constant, A'. The 'Bib. code' refers to the reference where these papers can be found if anyone should want more information. In finding the average, I have neglected the uncertainties and, where a range of values is given, I have taken the mean. We can see that the average of the values of the Hubble constant from twenty six of the latest measurements give a value equal to 'hr/m for the electron per cubic metre of space'. That is, the average value of H is equal to 'this much of an electron in each cubic metre of space' where 'this much of an electron is 'hr/m,' or just 'A'. Since the electron is not supposed to have anything to do with H at all, are you still convinced that the Universe is expanding?

Now coincidences do happen. The temperature of the Cosmic Microwave background radiation is 2.73 Kelvin. The number 2.73 is the freezing point of water in Kelvin (273K) divided by one hundred. This kind of coincidence is a 'think of a number, any number' kind of exercise and forms the foundations of numerology. What we have here with the Hubble constant, H is much more than numerology, very much more. The point is that h, r and m are not just 'any numbers', since we have met them before in our story of the Hubble constant.

A continuous spectrum of light consisting of all possible frequencies is emitted from the interior of stars, and the cooler gases surrounding the star absorb certain frequencies of this

Author	Bib. Code	Method	H Value
Cardone	2003acfp.conf..423C	Grav. lens	0.91A
Freedman.	2003dhst.symp..214F	HSTCepheids	1.1A
Tikhonov.	2002Ap...45...253T	HSt - Stars	1.2A
Garinge.	2002MNRAS.333..318G	X rays	0.89A
Tutui.	2001PASJ..53..701T	CO line T-F	0.94A
Freedman.	2001ApJ..553..47F	HST Cepheids	1.1A
Itoh et al.	2001AstHe.94.214I	X rays	0.94A
Jensen.	2001ApJ.550..503J	SBF	1.2A
Willick.	2001ApJ.548..564W	HST Cepheids	1.3A
Koopmans	2001PASA..18..179K	Grav. lens	(.94 – 1.1)A
Mauskopf	2000ApJ..538..505M	X rays	0.92A
Sakai	2000ApJ..529..698S	HST Cepheids	1.1A
Tanvir	1999MNRAS.310..175T	HST Cepheids	1.0A
Tripp	1999ApJ..525..209T	Ia Supernovae	0.97A
Jha	1999ApJS..125..73J	Ia Supernovae	1.0A
Suntzeff	1999AJ..117.1175S	Ia Supernovae	1.0A
Iwamoto	1999IAUS..183..681	Ia Supernovae	1.0A
Mason	1999PhDT...29M	X rays	1.1A
Schaefer	1998ApJ..509..80S	Ia Supernovae	0.86A
Jha	1998AAS..19310604J	Ia Supernovae	1.0A
Patural	1998A&A..339..671P	HIPPARCOS	0.94A
Wantanabe	1998ApJ..503..553W	Galaxies T-F	1.0A
Salaris	1998MNRAS..298..166S	TRGB	0.94A
Hughes	1998ApJ..501..1H	X rays	(.66 – .95)A
Cen	1998ApJ..498L..99C	X rays	(.94 – 1.3)A
Lauer	1998ApJ..449..577L	HST SBF	1.4A
		Average Value	**1.0A**

light. The frequencies of light absorbed by these cooler gases show up as black lines in the spectrum of light received from the star, and these dark lines are called 'absorption lines'. Electrons in the atoms forming the cooler gases surrounding the star are only allowed to have certain energies. If a photon comes along with exactly the right amount of energy to raise an electron from a lower energy level into a higher energy level then that photon is absorbed and the atom is said to be 'excited'. If the photon does not have exactly the right amount of energy to raise the electron to a higher energy level then the photon is not absorbed but continues on its way. When we look at the light arriving from distant galaxies we see that these absorption lines have been shifted towards the red end of the spectrum. This shift in wavelength was first interpreted as being due to a Doppler Effect, caused by the galaxy moving away from us and are now interpreted as a 'stretching' of space. However, we must always remember that in the first place, electrons in atoms surrounding the stars in the distant galaxies caused the absorption lines. This is why we must be surprised and very suspicious when we find that the value of the Hubble constant is related to a combination of the electron's mass and its classical radius.

Furthermore, as to which frequencies of light are absorbed depends upon their energy. If the energy of a photon of light is exactly equal to the difference between two energy levels for an electron in an atom, then the photon will be absorbed. The energy, E, of a photon is proportional to the frequency, f and the constant of proportionality is the Planck constant, h, so the absorption lines not only depend upon the electrons in the atom they also depend upon the energy of the photons of light and this, in turn, is dependent upon the Planck constant, h. As a consequence, Big Bang Codsmologists are telling us that the measured value of the Hubble constant is a combination of the things that they have used to determine its value in the first place!

In the Big Bang theory, H is not constant but varies as gravity slows the rate of the expansion down. One could argue that there must be a time when the two values, H and hr/m, are the same but why should that time be now, the first time that we measure it? This is why, when we see that H is equal to hr/m in each cubic metre of space, we must be extremely sceptical. As scientists we do not trust coincidences because they make us 'special'. True, we could all have been born at a special moment in time when H = hr/m - after all, someone has to be born then!

When coincidences like this happen, we usually have prior warning. Scientists predict that in the year such and such, the value of H will be hr/m. When this moment arrives, we all make a big party, celebrate, wish everyone a happy 'electron expansion' and get on with our lives. The point here is that at the first time we ever measure H with any degree of accuracy, it 'just happens' to have the same value as hr/m for the electron. In any case, to say that H varies with time is only speculation; we have no proof of this at all. It is only the Big Bang theory that says that the value of the Hubble constant has to vary – it may well turn out to be a constant after all.

Let us also remember that the age of the Universe or the 'Hubble time' is also determined from the Hubble constant. Since the age of the Universe is the inverse of the Hubble constant and since experiment tells us that the magnitude of the Hubble constant is equal to hr/m, if we are to believe the Big Bang theory, then we must believe that the age of the Universe is equal in magnitude to:

$$\text{Age of Universe} = m/hr$$

So not only could any school child take out a calculator and call up three very common constants and find the Hubble constant, by pressing the 'one over x' button, they could also determine the age of the Universe whenever they wanted! This sounds very strange indeed and yet if we are to believe codsmologists and the Big Bang theory then this is exactly what they are telling us. Does this mean that the electron is something special and related to the Universe itself? Is this why the Catholic Church celebrates Mass – the mass of the electron. I doubt it very much!

No, when we come across something like this, then we know that somewhere along the line, we went wrong. It is just like our car journey when we became lost. We thought that our navigation had been correct at every turn but when we found ourselves in the desolate farmyard, we knew beyond any doubt that somewhere on our journey, we had made a wrong turning. The fact that we ended up in the wrong place told us that we had gone wrong somewhere. It is the same with our vision of an expanding Universe. Nowhere in our Big Bang theory does the rate of the expansion of the Universe have anything to do with the electron or the energy of the photons of light - apart from the methods used in measuring it, and yet here we have H = hr/m in magnitude. We know that somewhere along the line we must have taken a wrong turn because of where we ended up and just as in our car journey, we have to go right back to the beginning and retrace every step to find out where we went wrong.

Could it be in the way we measure distances? Not really, as we now measure distances by many different means and have repeated the values many times. We measured the distance by finding the distance to the Hyades by statistical parallax but this is now done to a greater accuracy by the satellite Hipparcus that determines the actual parallax of the stars. This enables us to determine the distances to 'nearby' Cepheid variables, which in turn enables us to calibrate a distance scale using Cepheid variables and Supernovae Ia. The Hubble constant can be measured by other methods such as 'gravitational lensing' and in science, when one can measure the value of a physical quantity by two different methods and get the same value then this gives us even more confidence in our results. If a wrong turn was made then it was not in measuring distance. So, was our wrong turn made in determining the redshift?

It is unlikely, as energy levels within atoms are well known and can be measured with great precision, as can the shift in the wavelength of these absorption lines and so it is very unlikely that we went wrong in measuring the redshift. However, we must have gone wrong somewhere because it is asking far too much to believe that H is equal to 'hr/m' for the electron just by chance. In any case, the relationship between redshift and distance is both precise and repeatable and this is usually enough to tell us that what we are doing is correct.

If we have not made any mistakes in our experimental measurements then have we made any assumptions anywhere that could have been incorrect? Well, there was just the one. We had seen that stars, and nearby galaxies exhibited small redshifts due to their motion relative to the Earth and we put this down to a Doppler Effect. When distant galaxies exhibited huge redshifts we assumed that this was also due to the Doppler Effect and meant that they too were moving away from us – but with huge velocities. However, there were no intermediate values to measure! It was a big assumption in saying that since the Doppler Effect was responsible for the small redshifts measured in nearby stars and galaxies, it was also responsible for the gigantic redshifts found in the light from distant galaxies. Whilst we now talk in terms of space 'stretching,' the basic assumption is still the same. It is in this assumption that we have 'gone wrong'. We must look for a different effect that will give us a redshift that increases with distance, and knowing that $H = hr/m$ for the electron, we do not have to look too far for the culprit. Hubble always refused to call the shift in wavelengths in light travelling from distant galaxies 'velocities' because, in his words, "what we measure is redshifts and that is what I will call them".

What we need to look for is a mechanism by which the photons of light travelling to us from distant galaxies interact with the electrons in intergalactic space and thereby experience an increase in wavelength.

Chapter 14. Electron rules OK?

In the following chapters we are going to see how it really is. That is, just why are photons of light redshifted as they travel through space? Remember Fritz Zwicky and Tired Light? Well this is the modern version of that theory. In this theory we will see that photons are redshifted as they travel along because they interact with the electrons in intergalactic space. As they interact, they lose energy to the electrons and thus the frequency of the photon is reduced, the wavelength increased and they are redshifted. How do they lose energy to the electrons? We will see that as the photons travel along they are absorbed and re-emitted by the electrons in space with the result that the electron recoils each time. In this way, energy is lost from the photon to the recoiling electron; thus causing the photon to be redshifted. What happens to the energy lost to the electron? We will see that it is eventually re-radiated and it is this that gives us the CMB. Let's look at it all in more detail, starting with what a photon is and where the electrons come from.

One of the problems with considering redshift as the 'stretching' of a light wave due to space itself expanding is that we no longer believe that light is a wave. Gone are the days when we believed that light behaved in the same way as the continuous waves that we see at the seaside, since we now believe that light consists of particles called 'photons'. When you throw a stone into a duck pond, the circular ripples produced become larger and larger as the waves travel outward. As they travel outward, the radius of the circular ripples increases but their amplitude or 'height' reduces, showing that the energy of the ripple is becoming more and more spread out. With the photon or 'particle theory' of light, we see that the light energy comes in short bursts or quanta of energy called 'photons'. When we switch on the bedroom light, the room is not lit evenly by light waves spreading out from the bulb and 'lapping' gently on the walls. The room is lit by tiny 'pinpricks' of light illuminating various parts of the room, seemingly at random. Over a period of time, as more and more photons arrive, these 'pinpricks' build up to give an even illumination - in the same way as when one paints graffiti on a wall with a can of spray paint. With the spray paint, at first we see a blob here and then a blob there, but over a period of time the blobs arrive in such a fashion that we achieve an even coating of paint. It is the same with the light bulb in our bedroom. At first we get a speck of light here, then a speck of light there and over a period of time, it builds up into an even illumination. The reason this works, is that even the tiniest of light bulbs give out vast quantities of photons. For example, a sixty watt light bulb emits somewhere in the region of 10,000,000,000,000,000,000 photons each second so that in a normal bedroom this amounts to over two billion photons hitting each square millimetre of wall every second! It is little wonder that we 'see' these billions of photons as continuous stream of light.

The energy of each photon depends upon its frequency, the higher the frequency, the more energy each photon has. Since the colour of light depends upon the frequency too, different colours of light consist of photons of different frequencies or, to put it another way, photons of light having the same 'colour' all have the same frequency and thus the same energy. Photons of red light, which have a lower frequency than those of blue light, will each carry less energy. The energy of each photon is proportional to its frequency and the constant of proportionality is the Planck constant, h. The energy of each photon, E, is related to the frequency, f, of the light by the formula:

(Energy of photon, E) = (Planck constant, h)x(frequency of photon, f)

Or just simply;

$$E = hf$$

The idea behind the Tired Light theory of redshifts is that photons of light from distant galaxies travel through space for millions and millions of years and as they do so they will bump into the odd thing and lose some energy. According to our formula $E = hf$, if the photons lose some energy then their frequency must reduce. The same relationship between frequency, f, wavelength, λ and velocity, c applies for both waves and photons

$$c = f\lambda$$

With most things, if it were a car or a human runner we were talking about, as it lost more and more energy we would expect the speed to reduce as a result - but that cannot happen here. The speed of light in a vacuum or space is a constant and never varies. Einstein told us that and I am not going to argue with him. So, when the photon loses energy its frequency will reduce and, since the speed is constant, the wavelength must increase in order that our formula, $c = f\lambda$, holds true. The photon has undergone a redshift.

I always found this to be one of the hardest things to believe in the Big Bang theory. That is, no matter how far these photons of light have travelled, and some of them have been travelling for upwards of two hundred million years, they never ever lost any energy. That is, they never interacted with anything. To me, this goes against common sense. To say that a photon of light, a particle, can travel for millions of years and not lose any energy, just about goes against everything that I always knew to be true. With Tired Light it is just the opposite as here we expect light to lose some energy as it travels along. It is this loss of energy of the photons that makes the wavelength increase - this is how the Tired Light theory works!

Let us just pause for a moment to reflect.

From the supernovae results we saw that the measured value of the Hubble constant has the same magnitude as hr/m for the electron and we also saw that these constants were fundamental in the determination of galaxy redshifts. It is clear that we do not have to look far to find the particle that the photon bumps into on its journey through intergalactic space. Furthermore, the electron is responsible for the transmission of light through transparent materials such as glass and so we know that it interacts strongly with photons of light. What we are saying therefore is that photons of light travelling through intergalactic space (that is the space in between galaxies) bump into electrons and lose some energy on their way. This causes the frequency of the photon to reduce and its wavelength to increase. The photon is now nearer to the red end of the spectrum and so it has undergone a redshift.

Now doesn't that make much more sense? In the Big Bang theory, they say that photons from a galaxy twice as far away, experience twice the shift in wavelength. This is interpreted as the galaxy travelling away from us at twice the speed and due to the intervening space having stretched whilst the photon was in transit. In Tired Light, we say that light from distant galaxies travels twice as far through intergalactic space, bumps into twice as many electrons and therefore loses twice the energy. This gives us twice the redshift and this is what is measured, redshift not velocity. Velocity is just the interpretation that followers of the Big Bang put on the shift in wavelength. It never was the velocity that was measured - it was always the redshift. What we are saying is that redshift is caused by light losing energy in interactions with electrons as the photons travel through space – and this is why we have the coincidence between the Hubble constant, H and the parameters of the electron (hr/m) – Ashmore's Paradox.

To find out just how it does this, we need to consider how light travels through a transparent medium - even though Intergalactic Space is sparsely populated, it is still a 'transparent medium'. For this we need to consider Quantum Electro Dynamics or QED for short. In my eyes the expert in this field always was and always will be Richard Feynman. As an undergraduate Physics student, I like everyone else had my three volumes of "The Feynman Lectures in Physics" as, without these you just weren't a Physicist and never would be. Feynman, his diagrams, his QED and his alleged penchant for topless girlie bars made me what I am today and I try to emulate him in every way.

I well remember him talking on television about the meaning of 'light'. Is it a particle or a wave? He had no doubt about it - light is a particle. He likened the Universe to a swimming pool and imagined someone jumping in at the deep end. If light were a wave, as the person jumped into the pool the light waves carrying the information of the persons submersion, would spread out in the same way that the ripples of water would spread out - small, circular ripples getting bigger and bigger as they travelled. He then said; imagine a second person jumping into the pool somewhere else and then a third and a fourth. All these ripples spreading out from the different 'people sources' would overlap and, instead of them being easily identifiable separate waves, they would become confused as they crossed and overlapped and become just one big bewildering mess. Yet we can sort it all out instantly and know which person jumped in where. Feynman did not believe this at all. He felt that it was far more plausible that light consisted of particles or photons and that these photons travelled in straight lines direct from each person to our eyes. This he felt, was a far more convincing argument as to how the light travelled from one place to another rather than having millions of waves spreading out, interfering with one another and leaving it to the brain to sort them all out.

It is on this principle that QED is founded, that of photons of light travelling from one point to another. But what is a light wave? Basically, if an electron nearest to a light bulb starts to vibrate, then an electron slightly closer to you starts to vibrate followed by an electron closer still starting to vibrate, then we say that a light wave is travelling towards us. Light is a part of the family of waves that we call the 'electromagnetic spectrum'. These waves consist of vibrating electric and magnetic fields and they cause electrons in their path to oscillate or vibrate. We do not 'see' an electromagnetic wave; we just see the effect it has on the electrons within its path. This is how we transmit television signals; an oscillator at the transmitting station causes electrons to 'oscillate' up and down the transmitter antenna. This causes electromagnetic waves to be emitted and, when they arrive at the receiving antenna, these waves cause the electrons in the metal rods that comprise the antenna to oscillate up and down - hence the signal is detected.

Let's think about glass for a moment. When a photon of light travels through a piece of glass, the photon is constantly absorbed and re-emitted by the electrons within the atoms of that medium. In the spaces between the atoms the photon travels at the speed of light (3×10^8 m/s) but eventually our photon will bump into an electron and be absorbed. The energy of the photon is transferred to the electron and the electron within the atom starts to vibrate. What happens next depends upon the 'natural frequency' of oscillation of the electron in the atom.

The electrons in an atom have a preferred frequency at which they can oscillate and this corresponds to an energy level in the atom. If the frequency of the incoming photon is equal to the natural frequency of oscillation of the electron and atom then resonance occurs. The whole atom oscillates and the photon is not re-emitted. It has been absorbed (ultimately ending up as heat) and this is how our absorption lines in the spectra of stars and galaxies were formed.

If the frequency of the absorbed photon is not equal to the natural frequency of the electron within the atom then the oscillating electron re-radiates the energy as a new photon identical to the first. The photon continues on its way, in the same direction, until it bumps into another electron where it is absorbed and re-radiated. Our photon is constantly absorbed and re-radiated by the electrons as it travels through the glass.

However, the absorption and reemission of a photon takes time and so there is a delay between the photon being absorbed and the second 'new' photon being emitted. Or, to put it another way, when light travels through a medium, electrons in the atoms of the medium catch the photons, hop up and down for a bit and then emit a new photon identical to the first.

It is because of this delay between absorption and re-emission that the speed of light in a medium is less than it is in a vacuum. If one thinks about it, the more often a photon bumps into an electron within an atom, the more times it is absorbed and re-emitted and the more delays it suffers. The result of all these delays suffered at the hands of the electrons in the atoms is that it takes light longer to travel the same distance and so the average speed is reduced. As we said before, the photons travel at the speed of light between atoms but the result of all these delays as it is absorbed and re-emitted cause the overall speed through the glass to reduce.

A photon of light travelling through glass 'hops' from one electron to another in the same way that a frog crosses a lily pond. The frog hops onto one lily, rests, and then jumps onto the next lily rests and so on. In this way it crosses the pond hopping from one lily to the next. The only difference is that with the lily pond it is the same old frog that jumps off the lily leaf as that which jumped onto it in the first place. With light it is a completely new photon that is emitted by the electron, it just looks the same because it has exactly the same energy as the original photon that arrived.

However, there is another subtle difference. Whilst the electrons in a medium are generally firmly attached to the atoms in the glass, which in turn are firmly fixed into the glass itself, the lilies in the pond are only loosely fastened to the bottom of the pond by their roots. Unlike the electrons in glass, in our pond the lily leaves 'wobble'.

When the frog lands on the lily, some of the frog's energy is transferred to the lily leaf and the leaf plus frog move forwards. When the frog jumps off the lily, again some of the energy of the frog is transferred to the lily and the leaf will recoil backwards.

It is the same when a stationary gun fires a bullet, the bullet moves forwards and, in order that momentum is conserved, the gun must move backwards; this is called 'recoil'. Hence, as our photon of light travels through a chunk of glass and is constantly absorbed and re-emitted as it bumps into one electron after another we might expect that it would lose energy at each interaction. However, in glass this is not the case, as imagine our frog crossing a road. As it hops merrily along, the ground does not move because it is fixed. There is no recoil and so no energy is lost to the ground. Or, to put it another way, with our frog it is the whole planet Earth that recoils - but because it is so massive compared to the frog the earth does not move and we can neglect any recoil.

Coming back to photons of light travelling through glass, the atoms in a solid chunk of glass are one of many fixed into a crystal lattice and so here it is the whole crystal lattice that recoils. Since there are vast numbers of atoms in even the tiniest crystal lattice, then on an atomic scale, the mass is so huge that the crystal lattice does not recoil at all. If the intermolecular forces from nearby atoms restrain an atom then there is no recoil and no energy is lost to the atom when a photon is absorbed or re-emitted. Hence our photon travelling through the lump of glass loses no energy to the electrons because they are 'fixed' and cannot recoil. The energy remains constant and so the frequency and hence wavelength of the light remain the same. The light is not redshifted. This is a very important point in Tired Light theories. Any mechanism that is proposed for 'Tired Light' must be able to work in space but not work in glass or any other transparent medium. If it had the same effect in glass as it had in space then we would have a far greater redshift in the light as it passed through the optics in a telescope than it received in its entire journey through space. This is where the recoil interaction comes into its own. Densely packed atoms are unable to recoil and so no redshift occurs in these situations.

Let's apply this knowledge to photons of light travelling through intergalactic space. The space between galaxies is not solid like glass - otherwise nothing would be able to move about! But it is not 'empty' either, as most of Intergalactic space is made up of plasma. In fact more than 99% of all known matter in the Universe is in the plasma state. Plasma is a partially ionised gas. That is, a gas made up of positive and negatively charged particles formed when an electron is removed from an atom. The electron is able to move independently and is known as a negative ion whilst the remaining atom, which now has an overall positive charge, is known as a positive ion.

The word 'plasma' is a Greek word meaning "moulded' or 'shaped' and was coined by Langmuir and Tonks in 1929 who had been performing experiments on gas discharges. These experiments were the forerunner of fluorescent lamps and Langmuir was struck by the way the glow in the discharge tube took up the shape of the glass tube that contained it; hence the name, plasma. Plasma had been investigated even earlier by, amongst others, Crookes who called plasma the 'fourth state of matter," - the other three being solid, liquid and gas. However this is not a true way of labelling plasma.

In order to produce plasma there must be some mechanism by which ions can be produced from the neutral atoms in intergalactic space. If the ions are not constantly produced then the plasma will disappear, as the positive ions will eventually recombine with the negative ions to become neutral once more. As well as electrons being removed from the Hydrogen atoms to form a pair of ions, there is another mechanism by which ions can be produced.

In intergalactic space, cosmic rays smash into the nuclei of Hydrogen atoms and break them up. Cosmic rays are mainly protons - Hydrogen atoms with their single electron removed. They travel at almost the speed of light and are thought to originate in supernova explosions.

What cosmic rays are and where they come from was discovered at the beginning of the twentieth century. It was already known that a charged electroscope (a device for measuring electrical charge) gradually lost its charge no matter where it was placed – in the light or in the dark, in the open or in the middle of railway tunnels under tons of rock; the electroscope still lost its charge. It was thought that this effect was due to some sort of gamma radiation emitted from the ground. In 1910, Theodore Wulf decided to take a charged electroscope to Paris (he made the best electroscopes at that time) and climb up the Eiffel Tower with it. The point being that in passing through every eighty metres of air it was known that half the gamma rays emitted by the Earth would be absorbed and so the amount of gamma rays left over three hundred metres up at the top of the tower could be discounted. He found that the electroscope still lost its charge.

In 1912, Victor Hess took a charged electroscope up in a balloon to a height of five kilometres and found that the higher he went up in the balloon, the more quickly the electroscope lost its charge – meaning that whatever was discharging the electroscope was coming from up above and not from down below. At first it was thought that these cosmic rays were a part of the electromagnetic spectrum - hence the name 'rays' but now we know that they are not rays at all but consist solely of particles.

Intergalactic space contains Hydrogen atoms as they are everywhere in the Universe and are the basic building matter of everything we see. So when the proton in a cosmic ray whams into the proton that forms the nucleus of the Hydrogen atom the two particles are smashed into pieces called 'pions'. Pions in themselves are unstable and decay into neutrinos, positrons and yes you've guessed it, electrons. Hence we have plasma in intergalactic space.

The same process of cosmic rays crashing into the nucleus of atoms and producing pions occurs in our atmosphere as well but here, in the relatively densely occupied gas, the pions crash into other particles and lose their energy long before they have time to decay. Consequently, we get other particles as an end product and not electrons. The distance travelled by a particle before it decays is known as the 'decay length' and the distance travelled by a particle before it collides is known as their 'interaction length' or 'mean free path'. In intergalactic space, the interaction length is always longer than the decay length and so the pions always decay and produce electrons, positrons and neutrinos. Because of this, we can have more electrons in the void of intergalactic space than we have in our own atmosphere.

It is these electrons and positrons that make up the plasma of intergalactic space. The number of electrons in the space between galaxies is known to lie between 0.1 and 10 electrons in every cubic metre of space.

Whilst these charged particles are a long way apart they interact with each other by long-range electrostatic forces called 'Coulomb' forces. When we rub a balloon on a woollen jumper and place the balloon against a wall, it stays there. It is these same forces that attach the balloon to the wall that act between our electrons. Electrical forces between two charges are very strong but become weaker and weaker as the two charges are moved further and further apart. However, even with only one electron in every two cubic metres of space, the forces between several adjacent electrons are still strong enough to allow them to interact.

It must be remembered that overall, plasma is electrically neutral since it consists of equal numbers of positive and negative charges. In one way, these free charges are like the molecules in a gas in that they are not attached to any other particle in particular. They have large kinetic energies due to thermal effects and this gives the charges a random thermal motion. In fact the plasma in intergalactic space has a temperature between one hundred and one million Kelvin and this means that the average speed of the electrons in IG space is around one hundredth of the speed of light. This is fast, but not fast enough to bring in any significant relativistic effects and so we can treat the electrons as classical particles.

However, in other ways the motion of the charges is very different to that of molecules in a gas. Molecules in a gas are electrically neutral and so we can ignore any forces that may act between them - meaning that these particles are free to move as they please. In plasma, the charges exert forces on each other due to the electrostatic or 'Coulomb' forces acting between them. As we all learned at school, like charges repel, unlike charges attract and these forces act over very long distances indeed.

This means that, superimposed on top of the 'random' thermal motion, is a second motion determined by the electrostatic forces between the charges. That is, the random motion of the charges will be affected as the charges encounter each other and either repel or attract.

What is of interest to us is that charges in plasma can oscillate. That is, they can perform simple harmonic motion (SHM) and an electron that can do this can absorb and re-emit photons of light.

Let us look at the situation in dense plasma before we move on to that in intergalactic space. Whilst the charges are moving with their random thermal motion they tend to spread themselves out so that the plasma remains neutral. If an electron is displaced by some external effect then it will move towards electrons that are in front of it, thus increasing the repulsive forces acting on this electron due to the electrons ahead of it. The electron will also pass a few positive charges on the way - which means that there are now more positive charges behind the electron than there were before. Thus there will be a greater attractive force trying to pull the electron back. These forces combine and give rise to an overall restoring force trying to restore the electron back to its original position. When the electron returns to its 'original position' the electrostatic forces return to zero but the electron will have momentum that will carry it past this position and thus it will continue so that the electron is now displaced in the opposite direction than before. The direction of the restoring forces reverses so that they once again bring the electron back to its 'original position'. Notice that when we say 'original position' we are considering the electron to be initially at rest - which is not the quite the case in plasma but it is a valid simplification since the random thermal motion has no effect on the process. So, when we say 'returns to its original position' bear in mind that we are considering the two types of motion (random thermal motion and simple harmonic motion) separately.

Our electron therefore, is displaced to one side and the forces acting on it due to the other charges bring it back to its original position but its momentum carries it through this position so that it is now displaced on the other side. The restoring forces reverse and bring it back to the original position where, yet again its momentum carries it through. The electron in the plasma oscillates and the frequency at which it oscillates depends upon the plasma conditions – how many charges there are per cubic metre. The greater the density of electrons, the greater the restoring forces and so the higher the frequency at which the electron oscillates. This frequency is known as the 'natural frequency' of oscillation of the electron in the plasma.

Since an electron that can oscillate is able to absorb and emit radiation, the electron in the plasma is therefore able to absorb a photon of light. The photon comes along, bumps into an electron in the plasma and sets it into oscillation. The photon has been absorbed. As to what happens next depends upon the original frequency of the incoming photon.

If the frequency of the incoming photon has the same frequency as the natural frequency of oscillation of the electron in the plasma that absorbs it, then resonance occurs. The photon is absorbed and the electron set into oscillation. This then causes the rest of the electrons in the plasma to oscillate - all at the same frequency. The natural frequency of the electron is also the natural frequency of the whole plasma. The energy of the photon has now been dispersed amongst the whole plasma and so there is no way that the photon can be re-emitted.

If the frequency of the photon is much higher than the natural frequency of the electron and plasma then the photon is absorbed, the electron oscillates but this time resonance does not occur. The rest of the electrons in the plasma are not set into oscillation and the energy remains with the sole electron that absorbed it. The oscillating electron re-emits this energy as a new photon.

Think of your car when it has a front wheel 'off balance'. The wheel vibrates as it turns but as you speed up nothing much happens, until that is, you hit a certain speed. At this certain speed the whole car body starts to shake and rattle. Go above this speed and everything settles down once more. This is resonance. In order to give you a smooth ride, the car body has a suspension system which means that the massive car body is mounted on springs and can oscillate. Push down on the front wing of your car (when it is stopped!), let go and the car will oscillate up and down. The frequency with which it oscillates up and down is called its 'natural frequency'. Let's go back to driving the car along, when the frequency at which the rogue wheel vibrates is equal to the natural frequency of vibration of the car, resonance occurs and energy is transferred to the car body. As you maintain this speed, the wheel pulls on the suspension springs, the suspension springs pull on the car body - with the result that more and more energy is transferred to the car body from the wheel. The amplitude of the car body's oscillations becomes larger and larger until your car shakes itself to bits. At higher speeds and hence higher frequencies of vibration of the wheel, there is just not enough time for the car body to respond. The wheel pulls on the springs and squashes them but the car body has a large mass or inertia and so before the car body can respond, the wheel has changed direction and is going the other way. Work is done in squashing the spring and this energy is stored in the squashed spring. When the front wheel moves in the opposite direction, the spring does work on the wheel and so the wheel gets the energy back. Resonance does not occur, the wheel loses no energy and none is transferred to the car body at the other end of the springs.

It is the same with our photon and the plasma. Here the sole electron that absorbs the photon is the 'off balance wheel' and the rest of the electrons in the plasma represent the 'car body' on the end of the suspension springs. Electrostatic forces between the electron and the other charges in the plasma represent the 'suspension springs'. When the photon is absorbed, it sets the electron into oscillation with the same frequency as the original photon. This happens regardless of the frequency of the photon. If the photon has the same frequency as the natural frequency of oscillation of the plasma then the oscillating electron sets the whole plasma into oscillation and the energy is transferred from the electron to the whole plasma. The electron cannot get this energy back and so our photon has been absorbed, never to reappear. However if the frequency of our incoming photon is well above the natural frequency of oscillation of the plasma, then the photon is absorbed and the electron set into oscillation with the same frequency as the photon. This time however, the rest of the electrons are not set into oscillation and so no energy is 'lost' by the electron. It re-radiates the energy as a new photon.

The same happens in the plasma of intergalactic space (IG) even though the plasma is fairly sparse. We know the natural frequency of oscillation of this IG plasma is about thirty oscillations per second (30Hz). The frequency of photons of light is always well above this frequency (6×10^{14} Hz) and so the photons are always re-emitted. The energy is not transferred to the plasma; it is always re-emitted as a new photon.

However, the restoring forces that act between the electrons in the plasma of intergalactic space are not very strong since the electrons are far apart. They are strong enough to enable the electron to oscillate but not strong enough to hold the electrons firmly in place. That is the electron, just like the lily leaves on the pond when the frog landed on them, will recoil on absorption and re-emission of the photons.

Do we have experimental evidence of plasma absorbing and re-emitting photons? The answer is yes we do in the data from Pioneer 10. This is a space probe launched in 1972 and it was still sending signals until very recently. After flying past Mars and

Jupiter it has continued beyond the Solar system and out into space - on its way to the star Aldebaran where it will arrive in two or three million years. It has a gold plaque on the side with an invitation to aliens to come and visit us (let's hope they are friendly!) telling them what we look like, where we are in the Universe and so on. Whilst the signals from Pioneer 10 have posed a few problems for the scientists in that the spacecraft seems to have an unexplained acceleration towards the Sun (in other words, slowing down more than it should) they do tell us that that the plasma is absorbing and re-emitting the photons making up the radio transmissions. The distance to Pioneer 10 can be found by timing how long it takes for a signal to travel from Earth to the spacecraft and back again. To get the distance we just use the relationship between speed, distance and time - but we have to take into account that the waves travel slower than they do in a vacuum because of the intervening plasma. The photons have a frequency well above the natural frequency of oscillation of the electrons in the plasma and so they are absorbed and re-emitted. They are also subjected to a delay between absorption and re-emission hence the lowering of their overall speed. Consequently, there is evidence of plasma absorbing and re-emitting photons.

The electrons in the plasma of intergalactic space provide everything a Tired Light theory could wish for in life:

 i) They permeate the whole of space - so it does not matter from which direction the photons come ; they will all encounter similar amounts of electrons in travelling the same distance.

 ii) They are able to oscillate, which means that when a photon bumps into them, they absorb this photon and re-emit a new photon.

 iii) They recoil on absorption and re-emission, which means that some of the energy of the photon is transferred to the electron on both absorption and re-emission.

 iv) Since some of the energy of the photon has been transferred to the electron on both the absorption of the first photon and the emission of the second *new* photon, the energy of the second *new* photon will be less than that of the one absorbed in the first place. If the energy is less, the frequency is less and the wavelength longer. It has been redshifted.

The energy transferred to the electron on both the absorption of the original photon and on the emission of the new photon is relatively easy to calculate. Once we know how much energy has been lost to the electron on both the absorption and emission of the photon, we can find out just how much longer the wavelength of the new photon is, compared to the wavelength of the old one. When one does this, (the mathematics can be found on my website http://www.lyndonashmore.com) one finds the surprising result that all photons suffer exactly the same increase in wavelength regardless of their original wavelength. All photons have their wavelengths increased by an amount given by the Planck constant h, divided by both the mass, m of the electron and the velocity of light, c

Or to put it simply;

Increase in wavelength, $\delta\lambda = h/(mc)$

It does not matter if it is a photon of light that undergoes a recoil interaction with an electron or a photon of radio waves that undergoes a recoil interaction with the electron, the increase in wavelength will be just the same. After the collision the wavelength will be longer by an amount $h/(mc)$. But what happens to the energy transferred to the electron as it recoils; that is, the energy lost by the photon as it is absorbed and re-emitted?

This is important too. Let's go back to the lily pond and our old friend the frog, leapfrogging from leaf to leaf as it crosses the water. As the frog lands on a lily leaf the leaf, plus frog, will recoil slowly forwards with its newly found kinetic energy. Viscous drag forces or plain old 'friction' between the moving leaf and the water will bring it to rest. When the frog jumps off the leaf, the leaf will recoil in the backwards direction and plain old friction will slow it down again and bring the leaf to rest.

The energy transferred to the leaf ends up as heat in the water of the pond and is eventually radiated away.

It is similar with our photons and the plasma of intergalactic space. When a photon bumps into an electron and is absorbed, the electron recoils as it has received some of the photon's energy in the form of kinetic energy. The electron radiates this kinetic energy as a secondary photon and it is these secondary photons that form the CMB – and we will see dramatic confirmation of this when we calculate the wavelengths of these secondary photons and find that they are in the microwave region. Predicting the CMB is something that those who believed in the expanding Universe argued that 'Tired Light' Theories could not explain – how wrong they were!

Let us sum up how light loses energy as it travels through space.

Light consists of photons, and whenever light travels through a transparent medium it does so by being constantly absorbed by the electrons in that medium, which then re-emit a second *new* photon. The absorption of one photon and the emission of the new photon do not occur simultaneously, there is a delay whilst the electron oscillates up and down and during this delay, the electrons in space will recoil. In recoiling, some of the energy of the photon has been transferred to the electron and so the newly emitted photon has less energy than the one absorbed. Since the frequency of the photon is proportional to the energy, if the energy has been reduced the frequency is also less. The frequency of the photon is inversely proportional to the wavelength so if the frequency is less, the wavelength is longer. It has been redshifted.

The energy lost by the photon to the electron is radiated as a secondary low energy photon and this radiation forms the Cosmic Microwave Background radiation.

As the photons from distant galaxies travel through the plasma of space they will make collision after collision with electrons on their route and lose energy every time. As the energy of the photon gets less and less, the frequency gets less and less and the wavelength becomes longer and longer.

What Hubble and others have been measuring is the redshift or increase in wavelength of light as it travels from distant galaxies. They found that light from a galaxy twice as far away from us has twice the increase in the wavelength. A galaxy three times as far away from us has three times the increase in the wavelength and so on. This is, and always was, the Hubble Law - it never has been velocities. Interpreting the increase in wavelength, as a velocity was not an experimental result it was just someone's interpretation of these results to justify their theory of an expanding Universe. Now we know that this was the wrong interpretation.

What really happens is that photons of light from galaxies twice as far away, travel twice as far through the plasma of intergalactic space, bump into twice as many electrons and therefore lose twice the energy. The frequency of the photon is reduced by twice as much and its wavelength is increased by twice as much. Let's do that once more because it is an important result:

Light from galaxies twice as far away suffers twice the redshift because the light has travelled twice as far through space, made twice as many collisions, lost twice as much energy and suffered twice the increase in wavelength. Simple isn't it?

This is the real interpretation of the Hubble Law.

Chapter 15. Bull in a china shop.

So, we are saying that the explanation of the Hubble law is that photons of light from a galaxy twice as far away, travel twice as far through the plasma of intergalactic space, make twice as many collisions and thus undergo twice the shift in wavelength.

If only it were that simple!

As with all other stories, with 'the redshift in the light from distant galaxies' story, there is a twist in the tail. Granted, photons of light from galaxies twice as far away do experience twice the shift or extension in wavelength than those having only travelled half that distance but it is more complicated than that. The shift in the wavelength of the photon is also dependent upon the actual wavelength that the photon had in the first place. Redshift measurements show that the shift in wavelength, $\Delta\lambda$, is proportional to the original wavelength, λ such that the ratio of $\Delta\lambda/\lambda$ is a constant.

i.e.

$$\Delta\lambda/\lambda = \text{a constant.}$$

We call this constant the 'redshift' and give it the symbol, 'z'.

$$\text{redshift, } z = \Delta\lambda/\lambda$$

What this really means is that the greater the original wavelength of the photon, then the more its wavelength will be stretched as it travels through space. Red light has a greater wavelength than blue light so a photon of light at the red end of the spectrum undergoes a greater shift in wavelength than a photon of light at the blue end of the spectrum - even though the distance travelled by both types of light is the same. This results in the ratio 'increase in wavelength' to 'original wavelength' ($\Delta\lambda/\lambda$) being the same for all wavelengths. If a photon has twice the wavelength, λ, then it will experience twice the increase in wavelength, $\Delta\lambda$ so the ratio of $\Delta\lambda/\lambda$ will be the same.

Consequently, it does not matter which wavelength of light we measure – the redshift or 'z' for a particular galaxy will always be the same. Astrophysicists often use the value of the galaxy's redshift, z, to indicate just how far away the galaxy is.

For example, let us look at a galaxy at a redshift of z = 0.5. This means that all wavelengths are increased by a factor of 50%. Light with an original wavelength of 5×10^{-7}m when emitted would have a wavelength of 7.5×10^{-7}m on arrival. However, radio signals from this same galaxy having an original wavelength of 0.2m when emitted will have a wavelength of 0.3m on arrival at Earth. Whilst the actual increase in wavelength is different for different wavelengths the ratio ($z = \Delta\lambda/\lambda$) is always the same for a particular galaxy.

So not only is the shift in wavelength proportional to the distance that the photon has travelled, the shift in wavelength is also proportional to the original wavelength of the photon. Photons of light from galaxies twice as far away, travelling twice as far through the plasma of intergalactic space, and making twice as many collisions and thus undergoing twice the shift in wavelength, explains how the shift in wavelength is proportional to distance.

But how do we explain the relationship between the shift in the wavelength of a photon and its original wavelength?

To explain this result we need to look once again at how a photon travels through space. In the theory of Quantum Electro Dynamics (QED), as the photons travel along, they bump into electrons and are absorbed. There is a delay and then a new photon is re-emitted; some of the energy of the photon being transferred to the recoiling electron in the process.

However, the chance of a photon bumping into an electron in the plasma of intergalactic space, or anywhere else for that matter, differs from photon to photon. Whether a photon collides with an electron or not, depends upon the probability of it doing so and this probability is called the 'collision cross-section' of the photon. The collision cross-section has the symbol, σ (sigma) and it has different values for different photons. That is, some photons of light have a greater chance of bumping into an electron than other photons of light have. Collision cross-sections are a type of 'area' and indeed they have the units of area, the 'square metre'. We can think of collision cross sections as;

If the photon arrives within this area around an electron then the photon is absorbed. If this photon arrives outside this area around an electron, then it is not absorbed and the photon continues unaffected.

However, we must remember that the collision cross section is, in reality, a probability of the photon being absorbed. The process by which a photon is absorbed and its energy totally transferred to the electron, never to reappear again, is known as photoabsorption and the collision cross sections for this process are known from experiments using low energy X rays interacting with matter. Photons in the X ray region of the electromagnetic spectrum are identical to photons of radiation (including light) in any other part of the electromagnetic spectrum. It is just that they have a different frequency and wavelength. Consequently, what holds true for photons in the X ray region of the spectrum, should hold true for photons in other regions of the spectrum.

When a photon interacts with an atom in a medium the photoabsorption collision cross-section depends upon:

i) The classical radius of the electron, r (or to be precise, the diameter '2r').

ii) The wavelength of the photon, λ.

iii) How many electrons there are in the atom.

Or to put it simply, apart from the diameter of the electron, the bigger the wavelength of the photon or the more electrons in the atom, then the greater the chance of the photon being absorbed by that atom.

The formula for the collision cross-section for photoabsorption is given by:

$$\sigma = 2f\lambda r$$

Where the term 'f' is the 'atomic scattering factor' whose magnitude depends upon how many electrons there are in that particular atom and how much energy the incoming photons have. The factor 'f' is found by experiment.

In these experiments where X rays interact with matter, for Hydrogen (the simplest of all the atoms consisting of one proton and one electron) the values of 'f' vary between zero and unity and is dependent upon the photon energy. The value 'zero' indicates that the photon was absorbed and a new photon, identical to the first, was re-emitted so it looks like nothing happened to the photon. The value of 'one' means that the photon's energy matched an energy level in the atom and was completely absorbed. In this way 'f' is a probability – the probability that an atom will hang on to the photon once it has been absorbed.

The probability that the atom will absorb the photon in the first place is $2\lambda r$. Once the photon has been absorbed one of two things can happen, either the photon is retained or it is re-emitted. These two events are mutually exclusive and thus the sum of the two probabilities is 'one'. Since the atomic scattering factor, f, is the probability of the photon being retained, the probability of the photon being re-emitted must be $(1 - f)$. In our theory, we are looking at photons that are both absorbed AND then re-emitted and so to find the probability of the photon being absorbed in the first place and then re-emitted we must multiply the two probabilities together to give:

Probability of the photon being absorbed and re-emitted = $2(1 - f)\lambda r$.

For photons in the plasma of Intergalactic space, the frequencies of the incoming light are always well above the natural frequency of vibration of the plasma and so the atomic scattering factor, f, (the probability of it being retained) always has a value of zero. As a consequence, the collision cross section for a photon of light being absorbed and then re-emitted by an electron in the plasma is given by, $\sigma = 2\lambda r$.

Just for interest, let's see how the atomic scattering factor, f, varies for atoms other than Hydrogen. Beryllium has four electrons and the atomic scattering factor, f, varies between zero and four. Carbon has six electrons and here, 'f' varies between zero and six. Nitrogen has seven electrons and yes, you guessed it, the value of 'f' has values between zero and seven. This is hardly surprising since, to find the photoabsorption cross section, 'f' for the atom we simply add up the cross sections for all the electrons within that atom. This trend continues where the atomic scattering factor, f, has a maximum value equal to the number of electrons in the atom until we get to atoms with a large number of electrons, as here some of the electrons are 'hiding' behind others and so 'f' is a little less than the total number of electrons.

Returning to our photons travelling through the plasma of intergalactic space, the bigger the wavelength of the photon then the more likely the photon is to bump into an electron and be absorbed. This means that photons with twice the wavelength make twice as many collisions on their journey through intergalactic space and thus undergo twice the shift in wavelength than those photons with only half this wavelength. The ratio (shift in wavelength)/(original wavelength) is constant for all frequencies of photon.

The probability of the photon bumping into a particular electron and being absorbed is dependent upon its collision cross section and this in turn is dependent upon the classical radius of the electron and the wavelength of the photon. However the more electrons there are in the plasma of intergalactic space, then the greater the chance of the photon bumping into at least one of them.

Should the photon miss the first electron, then it will simply carry on until it bumps into the next electron or the one after that. It may be lucky and bump into the very first electron on its travels or it may be unlucky and have to travel a good long way, until it finally bumps into an electron. You just can't tell - but we would not expect this run of bad luck to go on for the whole of its journey. You win some, you lose some and it is the same with our photon. On its journey through intergalactic space, sometimes the photon only travels short distances between collisions with the electrons and other times, it will travel long distances between collisions with the electrons in the plasma of intergalactic space. However, the total distance travelled by the photon divided by the total number of collisions it encountered, will give us how far, on average, the photon went between collisions. This is known as the 'mean free path' of the photon.

The more electrons there are in the plasma of intergalactic space or the greater the wavelength of the incident photons, then the more times a photon will be absorbed and re-emitted by the electrons as the photon travels through space on its journey from a distant galaxy to Earth. The density of electrons in space is given the symbol 'n' and tells us how many electrons there are, on average, in each cubic metre of space. However, we must remember that if the number density of electrons becomes too high, then the photons will still make more collisions but they will not lose as much energy because the densely packed electrons will be restricted from recoiling. Redshifts only occur in sparsely populated plasma. If the number density becomes too high our redshift will disappear because the electrons will not recoil anymore. Whilst scientists do not know exactly how many electrons there are in intergalactic space, they have a good idea and it is estimated that, on average, there is between one and one hundred electrons in every ten cubic metres of space (n lies between 0.1 and 10 electrons per cubic metre).

The average distance between collisions or 'mean free path' can be found by:

$$\text{Mean free path} = 1/(n\sigma)$$

i.e.

$$\text{Mean free path of a photon} = 1/(2r\lambda n)$$

This relationship enables us to calculate how many collisions a photon makes on its journey through intergalactic space. All we have to do is divide the distance, d, between a distant galaxy and the Earth by the mean free path of the photon. In travelling from a galaxy a distance 'd' away, each photon of light of wavelength λ, will collide with an electron in the plasma of intergalactic space '$2dr\lambda n$' times.

$$\text{Number of collisions, } N = 2dr\lambda n$$

We have already seen that every time a photon bumps into an electron and is absorbed, the electron recoils. The electron then emits a new photon and recoils again. Each time the electron recoils some of the energy of the photon is transferred to the electron. Because the photon has lost energy, its frequency will be reduced with the consequence that the wavelength is increased. Furthermore, we have seen that the wavelength of the photon increases by an amount equal to h/mc on each interaction with an electron in intergalactic space.

So, if our photon makes $2dr\lambda n$ collisions on its journey through space and each time it interacts with an electron, its wavelength increases by an amount h/mc then the total increase in wavelength suffered by the photon in its entire journey must be 'how many times the photon collides multiplied by how much its wavelength increases each time' or, to state this algebraically:

Total increase in wavelength, $\Delta\lambda = (2dr\lambda n) \times (h/mc)$
or,

$$\Delta\lambda = 2ndhr\lambda/mc$$

The redshift, z is defined as the total shift in wavelength divided by the original wavelength or $z = \Delta\lambda/\lambda$. This tells us that;

$$z = \Delta\lambda/\lambda = 2ndhr/mc$$

The redshift or 'z' is the experimental result and we can see, shining like a beacon, the term 'hr/m' as mentioned in the paradox - but we will see more about this later. To obtain the Hubble constant we have to consider what the Big Bangers and the expanding Universe codsmologists did to this experimental result. They said that there had to be a velocity and, if there had to be a velocity, in the Big Bang, a velocity there would be!

In the Doppler Effect described earlier in the book, the velocity, v of a source of waves that produces a shift in wavelength of $\Delta\lambda$, in a wave emitted with a wavelength λ is given by $v = c\Delta\lambda/\lambda$ or the velocity of the source is equal to the speed of the wave, multiplied by the shift in the wavelength, all divided by the original wavelength. Since we call $\Delta\lambda/\lambda$ the redshift, z, what the Big Bangers call the velocity of the source, or in our case the galaxy, is just the speed of light multiplied by the redshift.

To compare our Tired Light result with the Big Bangers result gives:
$$v = c\,\Delta\lambda/\lambda = 2ndhr/m$$
or

$$\boxed{v = (2nhr/m)d}$$

Now, Hubble showed that the redshift, z of a galaxy was proportional to how far away that galaxy was. Big Banger codsmologists then converted the redshift into velocities and expressed the relationship as:
$$(\text{velocity of galaxy}) = (\text{Hubble constant}) \times (\text{distance})$$
or

$$\boxed{v = Hd}$$

Comparing this with our relationship readily shows that the Hubble constant is given by:

$$\boxed{\text{Hubble constant} = 2n(hr/m)}$$

When looked at in this way 'Ashmore's paradox,' where measured values of the Hubble constant actually equal (hr/m), is not a surprise at all. Since we know that there are between 0.1 and 10 electrons in each cubic metre of space it just means that the value of n is 0.5 electrons per cubic metre. However, taking our range of possible values of 'n', we can calculate a range of values within which the Hubble constant should, according to this theory, lie. That range of values is 0.41×10^{-18} s^{-1} and 41×10^{-18} s^{-1}. The generally accepted value for H - as found from experiment, is 2.4×10^{-18} s^{-1}. We can see that the theory is consistent with experimental results; the uncertainty is due to the lack of precision in determining just how many electrons there are in intergalactic space and not to any fault in the theory.

So we see that this Tired light theory can predict the value of the Hubble constant - but if only it were as simple as that!

We know that whenever a photon bumps into an electron in intergalactic space and interacts with this electron, the wavelength of the new photon emitted has its wavelength increased by an amount equal to h/mc. Since the wavelength has increased, the collision cross section has also increased and so the new photon is more likely to bump into the next electron.

The photon will not travel as far before it makes yet another collision; the mean free path will be shorter. As the photon travels through space and makes collision after collision, its wavelength will become longer and longer. The collision cross-section of the photon becomes larger and larger and so the photon makes more and more collisions as it approaches the end of its journey. The distances between collisions become shorter and shorter as the photons struggle to take the last few steps to arrive at the Earth. It is a little like skimming stones across a pond. The first few bounces are far apart but as the stone skims along the water surface, the distances between successive bounces gets shorter and shorter – decreasing rapidly towards the end.

There is a name for this and that name is 'exponential'. We meet exponential functions in many areas of science. The strange thing is that all these areas seem totally unrelated at first sight, and yet the end results or 'formulae' appear to be remarkably similar. Whilst the areas in science in which exponential functions appear are totally different, they all have one thing in common. That is, the rate at which something changes is proportional to how much of that 'something' you have. For instance, in nuclear physics the rate of decay of a radioactive sample depends upon how many atoms of that isotope there are in that sample. Let us say that there are one thousand atoms of a particular nuclear isotope in a sample and the chances of an atom decaying in a time period of one second, are one in ten. After one second, out of our one thousand atoms, one hundred atoms will decay in the first second, leaving us with nine hundred atoms in the sample. In the next second, one tenth, or ninety, atoms will decay thus leaving us with only eight hundred and ten atoms in our sample. In the third second, one tenth of these (eighty one) will decay leaving us with seven hundred and twenty nine atoms in our sample. In this way, we see how the decay rate decreases as we had one hundred atoms decaying in the first second, ninety in the second second, and a mere eighty-one in the third second.

The decay rate reduces exponentially. It is the same when a capacitor discharges. A capacitor is a useful electrical device that stores electrical charge. It is a bit like a battery that is running down or 'going flat'. When you connect a charged capacitor across a resistor, you have a big current at first but as time goes by, the current becomes less and less. The current reduces exponentially and this is because the rate at which the charge flows off a capacitor (current) depends upon how much charge there is on the capacitor at that time. Hence the current in our circuit reduces exponentially. In both these cases the quantity we are concerned with is reducing exponentially but there are cases were a quantity increases exponentially. An example of this is the spread of a contagious disease such as AIDS. With such a dread disease, the more people who have contracted AIDS, the more new cases there will be. If there are ten sexually active AIDS cases one night, then there will be twenty sexually active AIDS cases the next night (assuming each new partner contracts the disease) and forty the night after that and so on and so on. The number of new AIDS cases each day depends upon how many people have AIDS and did naughty things the night before. Or, to put this mathematically, the rate at which the number of new AIDS patients' increases is proportional to how many people had AIDS at that time. The number of people having contracted AIDS rises exponentially. The way to prevent the spread of contagious diseases is to put them into isolation (or use a little common sense!) so that the disease cannot be passed on. The trouble is, if left to their own devices, once the exponential function really kicks in, the number of new cases will increase with incredible speed!

To try to make the unimaginable imaginable I will tell you a little story. There are various versions of this story, which probably means it never happened, but if it did, the end result would be the same - the death of a peasant who thought he could outsmart a prince!

One day, a long time ago there was an Indian Prince who played chess against a peasant. To make it more interesting, they decided to make a little bet on the outcome. If the peasant won, then the prince would grant him anything he wished, whilst if the Prince won, then the peasant promised to take a bath and use deodorant before he played chess again. Well as you can imagine, after a great deal of excitement, cheering from the crowds and mass singing of football songs such as "You'll never walk alone", the Prince lost.

The peasant, on being asked to name his reward, instead of asking for the usual chest of gold or 'five of the Princes' 2000 wives' made a strange request. He asked that on the first day the Prince would give him one grain of rice and put it on the first square of the chessboard. On the second day the prince would give him double the previous amount and put two grains of rice on the second square of the chessboard. On the third day the Prince was to give the Peasant double the previous day's ration and put four grains of rice on the third square. This process was to repeat itself, that is, each day the peasant was to receive double the ration that he had had the day before until he had received a ration for each of the sixty-four squares on the chessboard.

This seemed a very modest request and so the Prince readily agreed to the Peasant's terms (but would have preferred for the peasant to have chosen five of his wives) and on day one gladly presented the peasant with one grain of rice. He was happier still on the second day when he presented the peasant with two grains of rice and offered the peasant more but the kind offer was turned down by the peasant on the basis that 'a bet is a bet!' Even on the eighth day when a whole row of the chessboard was accounted for and only 128 grains of rice were handed over, the prince was still happy at losing the bet, as he only had 56 more days to go and he had not even handed over enough grains of rice to make a whole meal. The peasant's wife was not too happy when she dreamt of what might have been.

"Why didn't you take the chest of gold like any normal peasant?" said she.

"Why didn't I take the five wives like any normal bloke," thought he.

But, no, a bet is a bet and that was the way it was.

But then the exponential function really kicked in, and an exponential function it truly was as each day's allocation of rice depended on how much he received the day before. On the sixteenth day, when two whole rows of the chessboard had been accounted for, he received 32,768 grains of rice - which was enough rice to feed his family and most of the street as well. On the 24th day when three rows of the chessboard had been accounted for, he brought home over eight million grains of rice and by the half way stage, he would have owned all of the rice in India had he lived to see it. As we said before, this is a fairy tale. In real life if you are a peasant and wish to stay alive, you either have to lose at chess or choose the five wives as a prize. As it was, the Prince (not wanting to go broke) shot the peasant for being far too clever. Had the peasant lived to collect the 64th day's ration it would have been 10,000,000,000,000,000,000 grains of rice, which is 10^{10} tonnes or the total World production of rice for the next 25 years!

This is an example of an exponential function. It increases by a fixed ratio each time. With our grains of rice on the chessboard, each day the number of rice grains doubles. We have one grain of rice on the first day, two on the second day, four on the third day, eight on the fourth day, and sixteen on the fifth day and so on.

With the photons travelling from a distant galaxy, the number of collisions they make also increases exponentially as they travel along. On each collision their wavelength becomes bigger resulting in their collision cross-section increasing. The rate at which they collide and therefore the rate at which the wavelength increases, depends upon the wavelength of the photon at that time. Hence we get the same exponential rise in the number of collisions as it travels along. Since the wavelength increases by h/mc each time the photon collides, if the number of collisions increases exponentially then the wavelength of the photon increases exponentially as it travels from a distant galaxy to the Earth.

The simple formula we had before now appears to become more complicated

$$z = \exp(2nhrd/mc) - 1$$

Where 'exp(2nhrd/mc)' means the exponential 'e' (e= 2.718..) raised to the power of (2nhrd/mc) – that is, we write e^x as exp(x) for simplicity. However, we have an expression for the Hubble constant, H.

$$H = 2nhr/m$$

This gives us the simple Hubble redshift relationship:

$$z = \exp(Hd/c) - 1$$

One might be forgiven for thinking that something does not tally here. The Hubble law v = Hd is a linear relationship and this Tired Light function is an exponential function. A linear relationship is one where you get a straight line passing through the origin, if one plots 'velocity', v against the distance, d. With a linear function such as v = Hd, the value of H, the Hubble constant must be a constant - hence the name, and this constant is the gradient or 'slope' of the graph of 'velocity', v against distance, d. Furthermore, if one doubles the distance the 'velocity' doubles, triple the distance, the 'velocity' triples and so on. This is in the very nature of a linear function.

However, this is the beauty of an exponential function. For small values the function is a perfect straight line. That is, the exponential Tired Light Hubble curve is linear for nearby galaxies. It only curves when the galaxies are far away. So if the distance - redshift relation is exponential why did Hubble and others find a linear relation when they plotted 'velocity' against distance? Remember the grains of rice and the chessboard? For the first eight days the amount of rice supplied rose at a pitiful rate, it was only later that it rose dramatically at a daily rate. In fact, if you look at day two, the peasant received exactly double the ration that he received on day one. This is a linear relationship. This is true of all exponential functions; when you draw a graph of the function it starts off as a straight line and only begins to curve upwards in a steeper and steeper curve in its later stages.

In the same way, for nearby galaxies the exponential function is, to all intents and purposes, a straight line and we obtain the same Hubble Law as Hubble did, v = Hd. But for galaxies further and further away, the exponential function truly kicks in and we must use the exponential form of the function.

Remember how the Big Bang theory had to introduce acceleration in order to explain how distant supernovae were further away than they should be? Well this is the real explanation. To measure distance they had used the apparent brightness of the supernovae. But when they compared this distance to the distance predicted by the redshift using a linear Hubble Law, the two distances did not agree. The supernovae were too dim for the distance predicted by their redshift! They said that the supernovae must be further away than expected and trumped up 'acceleration.' Well, we do not need to do this here, as what these supernovae are doing is to confirm an exponential Hubble diagram. An exponential Hubble diagram gives larger redshifts for the same distance and thus the results agree.

With this Tired Light theory, we are saying that the redshift increases with distance exponentially in the same way that the rations of rice increased exponentially on the chessboard. That is, the extra distance a galaxy has to be away from Earth for the redshift to double, gets less and less as it goes further and further away.

It has been known for some time that Hubble's diagram of redshift against distance was only a straight line up to a redshift of about 0.2. After this it took a turn upwards. Zwicky had proposed that the redshift distance relationship was exponential as early as 1929 but did not know why or how. Now we know both the why and the how's and we end up with an exponential relationship - but this in itself is not enough. We need some data and for this we need to look at the distant supernovae data as here we should see our exponential function in action.

In the Big Bang theory, high redshifts means high speeds and so, in the Big Bang theory, relativistic effects come into play. A redshift of '1' would show that the galaxy was travelling at the speed of light if relativity were not taken into account. Big Bang Codsmologists took the "Calan/Tololo Supernova survey" data and first 'corrected' this for relativistic effects due to their supposed high recession velocities. Having done this they decided that the data did not agree with the Big Bang theory's predictions. They decided that the results showed that the Hubble constant is greater now than it was in the past. In other words they decided that the results showed that the expansion of the Universe is accelerating!

In allowing for relativistic effects, they had assumed expansion and 'doctored' the data accordingly. The sceptics thought, what would the results look like if we didn't first assume expansion? That is, using this same data, what shape would the Hubble diagram have if we didn't 'doctor' it first for expansion?

In a paper published by Karim Khaidarov, he takes the same "Calan/Tololo Supernova survey" data as was used to show that the universe is 'accelerating' and shows that if it is not 'corrected' for relativistic effects, then it gives an exponential Hubble diagram. That is, supernovae data agrees with the exponential relation predicted by this Tired Light theory provided that we assume the universe to be at rest - that is, not 'expanding'.

Figure 1 shows the exponential Hubble diagram as predicted by this Tired Light theory. It can be seen that it is straight, or linear, for nearby galaxies, then, as the galaxies are further and further away, the curve takes an upward turn at around $z = 0.2$ and follows an exponential increase beyond this, as is shown by supernovae data when it is not 'corrected' for an expanding universe. This Tired Light theory agrees with experimental results. The Big Bang theory does not. When the data does not agree with its predictions it is put down to 'vacuum energy' or 'scalar fields' causing the expansion to accelerate. With Tired

Light, we do not need expansion, vacuum energy or the like; the data supports the predictions of the theory.

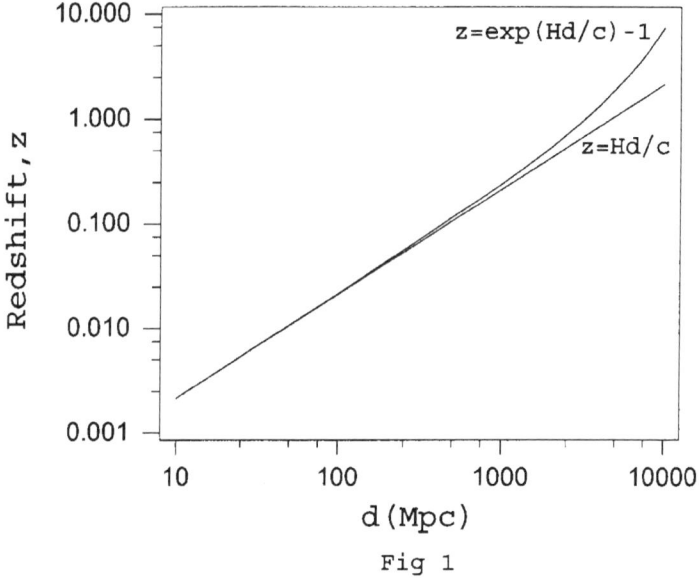

Fig 1

Chapter 16. Tired Light and the CMB.

One of the biggest sticking points of the Big Bang theory to those who do not blindly follow its preaching is the question "where does the energy go'? That is, whichever theory one follows, photons of light from distant galaxies have a longer wavelength on arrival and therefore a lower frequency than when these same photons set off. This means that the photons have less energy on arrival than when they set off. This energy must have been 'lost' on the way and so we must ask the question, "how did the photon lose this energy and just where did it go to?" How and where did the photon lose this energy?

In Doppler redshift, where the photons are stretched by the source rapidly receding, the question does not arise since the photon would never receive the energy in the first place. In the Doppler Effect, the photons experience a 'one off' stretch on emission due to a rapidly receding galaxy, and that would be it. It would continue on its long journey through space and be stretched no more. No matter where one detected it, either back here on Earth or in a spaceship close to the original galaxy, one would measure its wavelength as being the same (ignoring the speeds of either the earth or spaceship that is). Whilst this was the original interpretation of redshift, the Big Bang theory has moved on and the Doppler Effect is now only used to explain a small redshift imparted to the photons due to the star or galaxies' 'peculiar motion' that is, its motion relative to local space and not that due to expansion effects.

Cosmological redshift is said to be caused by the expansion of space. This is the one that Hubble and others measured and found to be proportional to distance. In the Big Bang theory, cosmological redshift is caused by space expanding and stretching the photons as they travel. The further away the galaxy, the longer it takes the photon to travel and so the more space itself will have stretched whilst the photon is on the way. Since space has stretched, the photon will also be stretched along with it and so the wavelength of the photon is longer. It has undergone a redshift.

The wavelength of the photon is longer, the frequency, and hence its energy must be less. But, why should the stretching of space cause the photon to lose energy? Where did the energy go?

Of course, there is the other little problem of wave particle duality. Photons are found to act like waves in some circumstances and like particles in others. The stretching of a wave is fairly easy to visualise but what about when the photon has its 'particle hat' on. How does the stretching of space explain the loss in energy of the photon in this case? Particles do not stretch and even if they did it would not necessarily cause it to lose energy. So just how do Big Bangers explain the loss in energy of the photons as they travel along?

To answer this question fairly we will use the words of a lecturer of cosmology whom I put this question to. I have condensed his answers for brevity, but basically this is our answer to the question.

"So, Prof. where does the energy go?"

"There are a number of ways of looking at this. One way is to say that the energy lost by the photons by being redshifted propels the expansion of the universe itself. Expanding the universe requires a great deal of work and this comes from the energy lost by the photons within it, CMB or otherwise. As the universe expands, masses move further apart and so the Gravitational Potential energy of the universe increases. The energy lost by the photons in being redshifted, CMB or otherwise, ends up as Gravitational Potential energy."

"That sounds good Prof., but if the expansion of the universe is being driven by the work done by the photons, which results in the photons losing energy and being redshifted, does this mean that the expansion has nothing to do with the original Big Bang? I understood that the effect of gravity was to slow the universe down and may well cause its future collapse into a 'Big Crunch!' If photons are providing the Gravitational Potential energy then there is no need for the expansion to slow down at all."

"True, in the inflation theory this would be the case and, since distant supernovae are dimmer than our theory says they should be, we say that the expansion is still accelerating."

"But, Professor, to do work one usually needs forces and the like, what 'universal' forces act on the photons to cause them to lose energy?"

"Let's try a different approach," said the Prof. "We have the General theory of relativity, which we use to describe the whole universe. In general relativity, space and time were created in the Big Bang at the same time as energy and matter. The principle of conservation of energy is sidelined in the General theory and the energy 'lost' by the photons has gone to curving space - time."

"But, Prof. we have yet to prove that space - time is indeed curved. Space itself is known to be almost flat. That only leaves the time component to be possibly curved. How do we know that that too will not turn out to be completely flat?"

"The General theory has thrown up many predictions that have been shown to be correct and so we expect it to be correct in this case. Space - time will be shown to be curved just wait and see."

So there you have the official party line on where the energy went.

As I said before, one of the biggest sticking points of the Big Bang theory to those who do not blindly follow its preaching is the question "where does the energy go?' The energy lost by the photons when they were redshifted due to the curvature of space - time, went to curving space - time, which caused them to lose it in the first place! No wonder they believe in curvature since the reasoning appears to go around in circles!

In the Tired Light theory, the photons lost energy as they travelled through space because of collisions between the photons and electrons in the plasma of intergalactic space. The result of this transfer of energy from the photons of light to the electrons in the plasma is that the electrons recoil and in doing so they accelerate. Whenever an electron accelerates it radiates energy and it is this secondary radiation that forms the CMB. The process by which the electrons radiate the energy is called 'Brehmsstrahlung' (the word 'bremsstrahlung' is German meaning 'braking radiation') and is caused by electrons accelerating inside an electrostatic field. In fact, Thomson showed that any accelerating electron will radiate energy regardless of whether it is near other charges or not. Thomson imagined the electron to be surrounded by its own electrostatic field with the field lines spreading out like the spokes of a bicycle wheel. Field lines show the direction of the force that would act if another charge were to be placed there. The fastest speed that information can travel is the speed of light and so, as our electron accelerates, it takes time for the information telling the outer reaches of the electron's field lines to move forwards and ensure that they keep up with the accelerating electron. A 'kink' travels outwards along the field lines and this distortion in the electrons own field causes forces to act on the electron itself with the resulting loss in energy being emitted sideways as a photon. A whole range of energies can be radiated and thus Bremsstrahlung is a 'continuous' form of radiation. That is, within reason, any frequency of radiation is possible (the exception, of course, is that the electron cannot lose more energy than it already has).

Thus the energy lost to the electron by the redshifted photon is then re-radiated and this secondary radiation forms the CMB. It is as simple as that, the principle of conservation of energy is upheld, there is no need for 'new' and undiscovered Physics such as curvature and what's more it makes sense! But let's look at this in more detail.

In Tired Light, if you remember, intergalactic space consists of plasma and we know how many electrons there are in each cubic metre of space. As a photon travels towards us it will be constantly absorbed and re-emitted by these electrons just as photons of light are absorbed and re-emitted as they pass through glass. The difference is that in glass the electrons do not recoil and so the energy of the new photon emitted is the same as the energy of the photon absorbed. There is no redshift in glass.

In space, the electrons in the plasma absorb the photon, which causes the electron to oscillate. The oscillations of the electron cause a new photon to be emitted. However, in space the electrons are few and far between and so they are not held rigidly in place. On absorption, the electron will not only oscillate but it will recoil as well. Thus, not all of the absorbed photon's energy is stored as the vibrational kinetic energy of the electron, but some of the photon's energy is transferred to kinetic energy of the recoiling electron.

Once the electron recoils within the plasma, forces (as described above) will act upon it to bring it to rest. The kinetic energy of the recoiling electron is emitted as a low energy secondary photon. These secondary photons, emitted as the recoiling electron is brought to rest, provide the CMB.

Having stopped recoiling, in order that light be transmitted, the oscillating electron re-emits a new photon - recoiling again, but in the opposite direction than before. The recoiling electron is again brought to rest by the process described above and a second secondary photon is emitted to add to the CMB. We can calculate how much energy is transferred to the recoiling electron - and thus the energy of the CMB photon emitted, and we find that it depends upon the wavelength of the incoming photon.

The relationship between the wavelength of the emitted CMB photon, λ_{CMB} and the wavelength of the incoming photon, λ can easily be calculated by the principle of conservation of momentum. The momentum of the incoming photon is h/λ and this momentum is transferred to the electron, which receives a boost in momentum of 'mv' when it recoils ('v' being its recoil velocity).

Consequently $h/\lambda = mv$ or $v = h/m\lambda$. Having found the recoil velocity we can calculate the gain in Kinetic energy of the electron using, $KE = mv^2/2$. This is the energy that is radiated as a secondary photon and the frequency of the CMB photon is found by using the equation, 'energy of photon = hf'. Having calculated the frequency, the wavelength of our CMB photon is found using '$c = f\lambda$' to give:

$$\lambda_{CMB} = 2m \lambda^2 c/h$$

A typical photon of light has a wavelength of 5×10^{-7} m, and if we substitute this into the above equation we see that the wavelength of the emitted CMB photon is 21cm, which is in the microwave region of the electro magnetic spectrum and interestingly, is the wavelength of the principal Hydrogen emission/ absorption line used extensively in radio astronomy. However, this is not the wavelength at which the CMB peaks. Experiment tells us that most CMB photons have a wavelength of 2.1 mm. To produce CMB photons of this wavelength requires the absorption and re-emission of an initial photon of wavelength 5×10^{-8} m. Photons of this wavelength are in the Ultra Violet part of the spectrum and there are an awful lot of these about in space.

We can see that the mathematics supports our view that CMB photons are produced from the energy lost by the redshifting of photons from distant galaxies. We do not need to put the principle of conservation of energy 'on hold' nor do we need to resort to 'curvature', an effect that has yet to be confirmed.

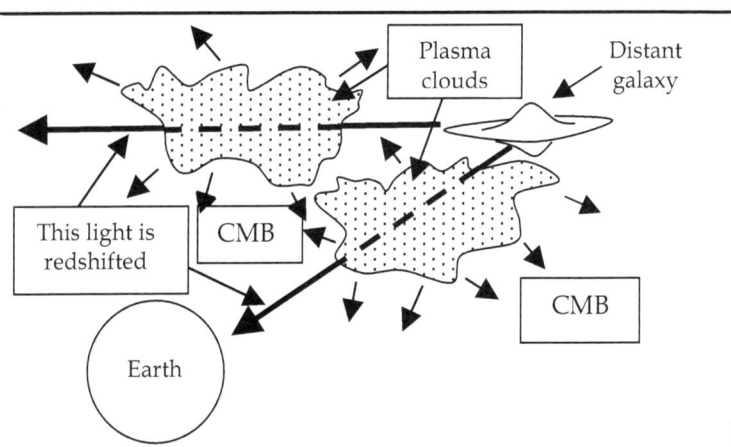

The distant galaxy gives off light in all directions. The light that travels directly to Earth is redshifted due to photons losing energy to electrons in the plasma clouds. The recoiling electrons in the plasma cloud radiate the energy lost to them as secondary photons – which form the CMB

In Tired Light, traditional Physics explains it all and the mathematics gives us wavelengths that agree with experiment. In fact, I have yet to receive a satisfactory answer as to why inflation is introduced to explain why the universe is so 'flat' and yet curvature is introduced to explain where the energy goes!

But, what of the Blackbody curve? Bremsstrahlung is a continuous process where any wavelength is possible. Unlike atoms, which emit discrete or line spectra, the deceleration of electrons produces a continuous spectrum. But Tired Light and the resulting CMB radiation answer another question. That is, in the past scientists have rejected the idea of space absorbing energy from radiation from distant galaxies on the basis that 'space' would heat up and eventually 'glow'. Well, in a way, with this theory this is exactly what happens – the plasma clouds absorb energy from the photons as they are redshifted and the plasma clouds 'glow' – but the 'glow' is not in the visible region, it is in the microwave region. The plasma clouds are in a state of thermal equilibrium where they radiate energy in the microwave region at the same rate as they receive it from the redshifted photons of light passing through the plasma clouds. Why does the intensity of the CMB radiation peak at a wavelength of 2.1mm? Well, as we have seen, CMB photons of this wavelength are produced by redshifted photons in the UV part of the spectrum. Redshifted photons originally with lower energy than these give off less energetic CMB photons so the intensity of the CMB falls off at lower frequencies. Whilst redshifted photons originally with higher energy than these give off more energetic CMB photons, there are fewer and fewer of them and so again the intensity of the CMB falls off at higher frequencies. Thus we have accounted for the Hubble law and the CMB with our Tired Light theory.

Explaining the CMB in this way answers another question which is "Just why are there so many CMB photons?" There are far more CMB photons in the Universe than any other, so why? Well, a photon of light on its way to us from a distant galaxy will be absorbed and re-emitted many times on its journey to us. In fact, a single UV photon of wavelength 5×10^{-8} m (the wavelength that produces CMB photons at the peak of the CMB intensity curve) will produce over 40,000 CMB photons in undergoing a redshift of just one ($z = 1$). This theory actually predicts that there will be many more CMB photons than any other.

Another success the theory has is in explaining the 'clumpiness' of the CMB. Whenever a photon is redshifted by the absorption and re-emission of a photon the electron recoils twice and two CMB photons are given out. For photons redshifted within these plasma clouds, the clouds will 'glow' in the microwave region and provide the CMB. Whilst the law of averages tells us that the microwave radiation given off by each cloud will be very similar there will be slight 'temperature' variations from cloud to cloud. These will appear as clumps in the CMB. These clumps are known to exist but, in the Big Bang theory they are said to be the 'seeds' from which the galaxies etc grew. In the BB Theory, these clumps are said to be at the very edge of the Universe and seen when we look back in time almost at the Big Bang itself.

It would seem to be an easy dispute to settle. In the BB, these clumps are very distant. In Tired Light the clumps in the CMB are 'local' and formed by nearby plasma. If the clumps are distant, as in the BB, then there should be no relationship between our galaxy and the 'clumps' in the CMB as this would put us somewhere 'special' in the Universe.

In 2005, a team of international scientists found that the larger of the clumps in the CMB were aligned with our local galaxy cluster and some are actually arranged in a regular pattern around the plane of our own Solar System! In astronomy, large means close and so we are not surprised when we find that it is the larger or closer of the clumps that are aligned with our local galaxy cluster. Furthermore, galaxies such as ours are known to be surrounded by hydrogen clouds and these are thought to be the remnants left over from the formation of our local galaxy cluster. The Tired Light theory says that cosmic rays in the form of protons passing through these Hydrogen clouds collide with the Hydrogen Nuclei and create plasma. Photons of light passing through the plasma are redshifted and the energy lost is re-radiated as the CMB. The clouds give us the 'clumpiness' and since these clouds are the 'left overs' from the formation of the galaxies in our local cluster, is it any wonder that the larger and nearer of the clouds are linked to the galactic plane of the Milky Way?

Further evidence for this exists in reports that the redshifts of distant galaxies are actually quantized! The '2df redshift survey' was designed to look at the redshifts of around 250,000 galaxies and what some researchers claim to have found is that whilst there is an even spread of galaxy redshifts there are certain redshifts much more popular than others. Furthermore, these redshifts occur in certain multiples of a certain, basic quantity. Just as Department chain stores are quantized in that one has one store, two stores, three stores etc but never 'one and a half stores,' the redshifts in the galaxies came in multiples of around $z = 0.025$. In the BB theory, this means that after a huge explosion at the beginning, the galaxies end up regularly spaced throughout the universe – like rings in an onion, but centered on Earth! Tired Light would say that if the galaxy redshifts are quantized then it is caused by plasma clouds - and the light is travelling through one cloud, two clouds, three clouds on its way to us and that it is the plasma clouds that are evenly spread throughout the universe and not the galaxies.

If the alignment of the larger of the clumps in the CMB with our galaxy is not enough to convince you that the CMB is local and not distant as in the Big Bang theory, then lets look at 'gravitational lensing'. Gravitational Lensing is predicted by the General Theory of relativity and occurs when a large massive object bends the light passing by it and 'magnifies' distant objects behind it. It is known to exist and appears quite a lot with the study of quasars. Since the CMB extends throughout the Universe then, in the Big Bang theory, it is predicted that gravitational lensing should have an effect on the CMB. However, it does not! There is no known case where the CMB is found to be gravitationally lensed - thus showing that the CMB is not distant.

Furthermore, if, as in the BB Theory, the CMB has nothing to do with the plasma clouds, then there is some interesting correlation between the temperature of the plasma in IG space and the CMB which would have to be put down to coincidences. Let's have a look at what the BB calls Heavenly happenstance!

Firstly there is the "The 'coincidence' between the plasma temperature and the wavelength at which the intensity of the CMB curve peaks." We know that the temperature of the plasma clouds in intergalctic space is between 10^5 and 10^6 Kelvin and we know that the average kinetic energy of an electron at these temperatures is given by $3kT/2$ - where k is the Boltzmann constant (1.38×10^{-23} JK^{-1}). The point where the energy of the incoming photon is equal to the initial kinetic energy of the electron that it interacts with is of interest and should mark a watershed in the CMB curve. If the energy of the incoming photon is much less than the kinetic energy of the electron then the electron's motion will not change drastically but will only be modulated by the effects of photon absorption. If the energy of the incoming photon is much greater than the initial kinetic energy of the electron then the effects of photon absorption will dominate. The point where the photon energy is equal to the electron's kinetic energy marks the point where one effect finishes and the other starts. We will now work out the wavelengths of the CMB photons emitted at this watershed and see if it shows up in the observed CMB curve.

> Energy of incoming photon = hf
>
> Average KE of plasma electron = $mv^2/2$ = $3kT/2$.

For this temperature range (10^5 to 10^6 Kelvin) the range of frequencies of incoming photons having the same energy as the average Kinetic Energy of the electrons in the plasma is 3.1×10^{15} Hz and 3.1×10^{16} Hz. When these photons are redshifted by Tired Light, the wavelength of the CMB photons given off will be in the range 0.076mm to 7.6mm - and it is in this range of wavelengths where the intensity of the CMB curve should peak. Again, this is consistent with observation as the wavelength at which the intensity of the CMB peaks is found to be just that – 2.1mm! The uncertainty is due to the uncertainty in determining the temperature of the plasma clouds. However, here we have introduced a new parameter - the temperature of the plasma clouds, and yet the Tired Light theory is still consistent with observation. Below is shown the CMB intensity curve and we see that the wavelengths lie within our predicted range of 0.076mm to 7.6mm.

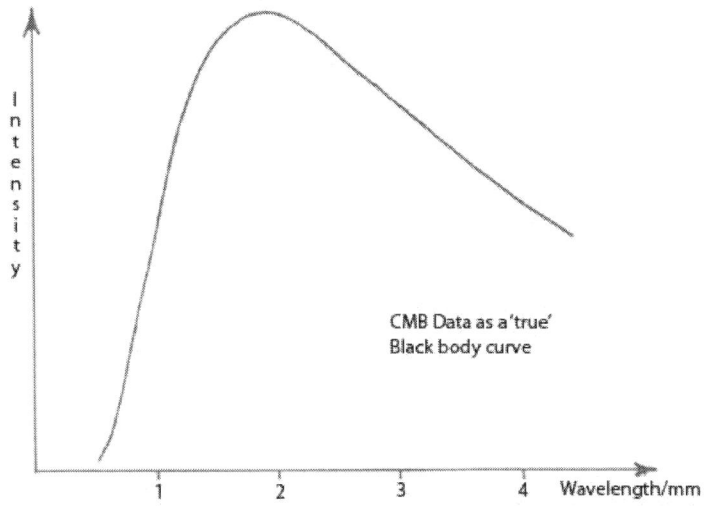

Notice that this curve has been drawn the 'correct' way around. Big Bangers usually draw this curve as intensity against frequency, or use a non linear scale for the wavelength and some even use 'waves per cm' on the bottom axes instead of wavelength – I can't think why, but the resulting curve that they get looks much more like the traditional Black body curve when one does it this way. I wonder if that has anything to do with it! Always look carefully at the axes when looking at these graphs. The two curves below look alike don't they? They are the same shape and convincingly show us that the CMB curves are really Blackbody. But do they? When one looks at the bottom scales,

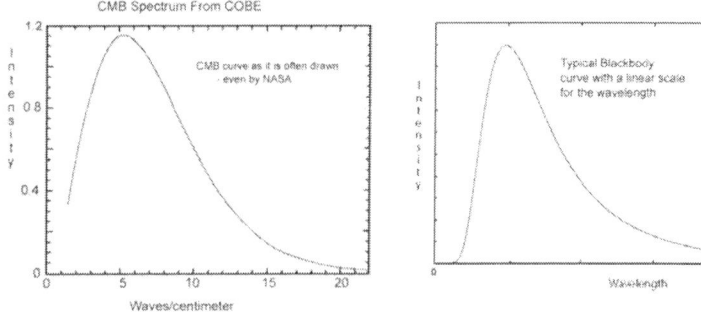

we see that they are very different. The typical blackbody curve on the right has a horizontal scale linear in wavelength - just as it should have. However, the curve on the left shows the CMB spectrum from the COBE satellite and FIRAS data. This has been drawn by Big Bangers and has 'waves per centimetre' along the horizontal axis! This is a very different kettle of fish as this tells us not the 'wavelength in centimetre' but 'how many waves there are in each centimetre.' Consequently it is not the wavelength as it should be but 'frequency' in disguise. Even NASA websites present the information in this way. Compare the typical blackbody curve with the 'true' CMB curve shown earlier and it is not as convincing is it? Now I wonder why they do this?

Returning to our Tired Light theory, do you remember the 'coincidence' of how the CMB curve is related to the range of temperatures in the plasma clouds? Well we also have "The 'coincidence' between the point at which Compton Scatter ceases and the wavelength at which the intensity of the CMB curve peaks." There are other interactions between photons and electrons and one of the more common ones is the Compton Effect. Here, the photon is scattered at an angle to the original direction in which the photon was moving. There is an exchange in energy between the photon and the electron but the photon is scattered. It does not arrive at Earth since the direction of the photon has changed. In the past, this effect has been put forward as a possible cause of redshift but it fails because the photons have to undergo a change in direction in order to lose energy. This means that the photons would be scattered off in all directions on their way to us and we would not see anything in space clearly. This Tired Light theory uses the same mechanism as that used by light as it travels through glass and so 'scatter' is not a problem here as the photons are absorbed and re-emitted along the same straight line.

As to 'which way the energy goes' in the Compton Effect depends upon the energy of the photon compared to that of the electron. In the Compton Effect, if the photon has more energy than the electron then the electron gains energy and the photon loses it. This is what is normally known as 'Compton Scatter'. However, if the energy of the photon is less than the energy of the electron then the photon gains energy from the electron. The electron slows down and the photon gets a boost in frequency. This is known as 'Inverse Compton Scatter'.

When the energy of the photon is equal to the energy of the electron it collides with then there is no exchange of energy by the Compton effect at all. At these photon energies, there is no Compton Effect! As we have shown above, this is the point in the Tired Light theory where virtually the whole of the CMB spectrum is formed! Photons with energies equal to the average kinetic energy of the electrons they interact with produce the microwave CMB - and yet this is the very same point at which Compton scatter is no longer having an effect. Is it because the two effects, Compton and Tired Light, are competing for electrons to collide with? At the point where the photon energy is equal to the kinetic energy of the electron in the plasma cloud, there is no exchange of energy between the photon and electron fields by the Compton Effect. Tired Light Effects dominate.

As we have seen with 'Ashmore's Paradox', measured values of the Hubble constant are equal to 'hr/m for the electron in each cubic metre of space'. In the BB, this is a coincidence! When we calculate the wavelengths of the secondary photons given off when photons lose energy as they are redshifted by Tired Light we find that these secondary photons are in the microwave region. In the BB, this is a coincidence! Here we introduce the temperature of the plasma clouds and we find even more coincidences. The range of CMB photons is produced by photons having the same energies as the average kinetic energy of the electrons in the plasma of IG space. In the BB this is a coincidence! The range of CMB photons is produced at just the point where the Compton Effect stops and has no effect. In the BB this is a coincidence! For me there are just too many coincidences going on here between redshift, CMB and the electrons in the plasma of intergalactic space. As far as I am concerned, these things are related! The Big Bang theory and the expanding universe are just too big a stretch of the imagination!

Chapter 17. Supernovae time dilation.

From the ancient Greeks to modern times, scientists have always been sure that they have understood the workings of the universe. It therefore comes as no surprise that scientists have been sure that the universe started in a Big Bang and has been expanding ever since. Being sure is one thing, finding proof is another. Penzias and Wilson's discovery provided direct evidence that there is a background radiation – a CMB, but this did not prove that the universe was expanding; it only proved earlier steady state theories wrong (but not later ones) and certainly did not prove Tired Light models wrong. Redshift itself does not prove the Big Bang theory correct as Tired Light explains this phenomenon and also predicts a value for the Hubble constant that is consistent with observation. The ratio of Helium to Hydrogen atoms in the universe being around 1 to 4 does not prove the Big Bang correct as Helium is also produced from Hydrogen in stars in a similar ratio. During the nineteen seventies and eighties, whilst we were all being assured that scientists understood everything and the universe was *known* to be expanding there was still no direct proof of this. It was still just a theory - only an idea.

But in the laboratory's backroom, they frantically looked for ways that would provide the proof that they not only craved, but needed. They looked for something that stank of expansion. In an attempt to correct this situation, astrophysicists needed something that only an expanding universe could do and a static one could not. What they came up with was the 'Tolman surface brightness test' and 'supernovae time dilation' – let's look at time dilation first.

Here they used the same supernovae as they used to determine the Hubble constant but this time they looked for supernovae a long, long way away so that their huge galactic recession speeds would display relativistic effects. An expanding Universe will involve speeds approaching the speed of light and hence relativistic effects will predict time dilation. Time dilation is where 'time clocks' appear to run slowly and thus cause things to age more slowly. Imagine a digital clock on your wall. As time slowly ticks away, the numbers displaying the 'seconds' will count in a regular way. Now imagine the same clock on a distant but stationary star. It takes time for the light carrying the information to arrive at our eyes. What we see now is the clock appearing to be 'slow' since we are seeing it as it was when the light was, emitted and not as it is now. However, the seconds will tick away in the same regular manner and the time interval between one second and the next will just be the same as it was when the clock was on our wall. Now imagine our clock on a distant supernova moving away from us at almost the speed of light. At the 'top of the hour' the light carrying this information sets off and travels towards us. During the next second the supernova moves backwards so that when the digit denoting that one second has passed appears, the light carrying this information has to travel further than the previous one and so takes longer to arrive here on Earth. We see not only that the clock is slow but we see each second as being longer than here on Earth. This is 'Time Dilation' and the closer the supernova travels to the speed of light, the more time is 'stretched' or 'dilated'.

In the Big Bang theory, the further away the supernova is, the faster its speed of recession will be, and relativistic effects will cause us to see it decay more slowly. As a consequence, in an expanding Universe, the further away the supernova, the longer it will take to decay. In a 'static universe', galaxies will not have the huge speeds and will not involve time dilation. In this case, all supernovae will decay at the same rate, regardless of how far away they are.

In the thermo-nuclear bomb that is the supernova, heavier and heavier metals are fused together as the star collapses. As a consequence, the chemical composition and hence emission spectra, of the fireball changes with time. There are no emission lines of Hydrogen present since all the Hydrogen was burned up long ago, before the star collapsed. In the dramatic death throes of the star, the rapid rise to a maximum brightness is due to nuclear fusion, but after this, the supernova is kept alive by the decay of the isotopes produced by the original 'bang'. This is why the supernova gradually disappears over a period of a year or so. An unstable isotope of nickel decays to cobalt, which decays to good old stable iron as the supernova fades away. The rate at which these isotopes decay is determined by their half lives and these are a constant even for a supernova. By observing the tail of the light curve through a spectroscope, one can see the spectrum change, as different elements pass through the outer atmosphere (called the photosphere) on any particular day. This gives us an excellent clock with which to measure the decay. Whilst the half-life of a radioactive element is a constant for a stationary supernova, relativistic effects will come into play for a distant one travelling away from us at almost the speed of light. The further away the supernova, the nearer its recession speed approaches the speed of light and the slower the isotopes will be seen to decay and change from one form to another. We see this as the more distant supernova taking longer to fade away.

Scientists divide the spectrum of the supernova into several small 'features', with each feature denoting a small range of wavelengths. They then measure the time delay between certain wavelengths of light arriving – these certain wavelengths corresponding to the emission lines emitted by the isotopes, as one decays into another. Now, it must be said that some cast doubt that the experimental results show time dilation at all. The sample of supernovae is small and one has to be careful of the Malmquist bias.

The Malmquist bias is a sampling error in the statistics. In order to look for time dilation in the supernova light curves we must look at supernova further and further away so that the relativistic effects become significant. In other words we need supernova with a redshift close to 'one' or above so that the recession 'velocity' is approaching the speed of light. As we look further and further away, the effects of dust between us and the supernova comes into play, and this, added to the effects of the inverse square law have the combined effect of reducing the apparent brightness of the distant supernova. As a consequence, we are unable to see the dimmer supernovae, we only see the brightest ones and this is where the Malmquist bias can rear its ugly head. For measurements on supernovae at the required distance away, we cannot use 'ordinary' supernova as they are too dim to be seen. That is, we are not using the normal 'man in the street' supernova that statisticians usually go to great lengths to find, but we are using the biggest and brightest supernovae ever seen. To put it another way, the conclusions of time dilation are based upon data, which is based upon a small sample of freaks! The supernova samples are not normal. The further out we go, the brighter the supernovae we use for our measurements and thus we are not using a consistent set of supernovae for our data. The question is; is this a 'fair test' or more importantly, do the results have any meaning? Is it that brighter supernova just take longer to decay?

I will leave that particular question for the reader to ponder upon and continue assuming that the results are valid. That is, we will assume that 'time dilation' exists. Big Bangers tell us that in a static universe, there will be no time dilation whilst in an expanding Universe, time dilation will occur and be proportional to $(1 + z)$. In other words, the light curve of a supernova with a redshift of one will take forty days to reach maximum brightness instead of the twenty days of one 'close by'. They claim that the results show time dilation. However,

Quasar light curves tell a different story. In the 1960's, radio telescopes detected intense radio sources and so astronomers naturally turned their optical telescopes onto these objects to see what they looked like. They saw various things but nothing special; in many cases it was just a distant galaxy or an ordinary star. Because they were a star-like radio source they called them Quasi-Stellar Radio Sources or Quasars for short. As time went on, they found other, similar objects, but these were not strong emitters of radio waves and these were called Quasistellar Objects or QSO's for even shorter but the term 'Quasar' sounds interesting so it stuck even though 99% of Quasars do not emit any radio waves!

Quasars are very bright and when astrophysicists looked at the absorption lines in the spectra from the Quasars they were baffled since they didn't look like anything that they were used to. It was then that they realised that the reason for their quandary was that they were looking in the wrong place! The

Quasars had such an enormous redshift that the absorption lines were not just shifted, but they were in a totally different place in the spectrum altogether. Quasars with redshifts of up to five and beyond are not uncommon and this puts them at huge distances away from us - which raises a problem, as normally the further away something is, the dimmer it is; yet here we have huge distances and very bright objects. If Quasars are at the huge distances suggested by their redshifts, then their true brightness must be unimaginable - and certainly places them amongst the very brightest objects in the Universe. There is still a great deal to be discovered about Quasars but they appear to be linked to active galaxies containing super massive black holes.

Any story of redshifts and Quasars is incomplete without reference to Halton Arp. Halton Arp was born in New York and studied at both Harvard and the California Institute of Technology, going on to conduct research at the Mount Palomar Observatory. At one time, he worked as Hubble's assistant but soon found himself at odds with the scientific community. Arp's failing was that he consistently made observations that went against the Big Bang theory and, instead of keeping quiet about these discoveries; he reported them to the embarrassment of his colleagues. Arp claimed that Quasars were consistently found not just close to a particular type of galaxy, but were symmetrically placed at either side of that galaxy - showing that there was a link between the two. His colleagues doubted his findings and questioned the statistics so Arp went on to put the matter beyond doubt by compiling a catalogue of over four hundred of these pairs, including one picture showing a 'bridge' of material between two objects having completely different redshifts. Since science is based on scientific evidence, one might have thought that this was enough to convince even the most hard-nosed sceptic, but it wasn't. His colleagues at both the Mount Wilson and Palomar observatories are alleged to have gone to the directors of the observatories and suggested that he be banned from using the telescopes, thereby stopping the flow of experimental results that were detrimental to the Big Bang theory! The Directors are said to have agreed to the ban, with no amount of reasons or pleas changing their minds. This had the end result of Arp moving to Germany, where he remains a thorn in the side of the Big Bangers and is still active and writing books and papers that are in conflict with the Big Bang theory. However, it must be said that for someone who is claimed to have been kept silent by mainstream science, he seems to have achieved a great deal of publicity and had many papers and books published - with the result that everyone has heard of him and his theories!

Arp believes that the Quasars are not far away in the distance but that they are relatively nearby. He, and others, believe that quasars are newly created matter thrown out of the active galaxies, with their huge redshifts being 'intrinsic' and due to some property of the quasars themselves. That is, a Quasar's redshift is partly due to cosmological causes (expansion) but mainly due to something happening within the Quasar itself. Big Bangers insist that Quasars are gravitationally 'lensed' by the galaxy. The massive, nearby galaxy 'curves' the space in the region around it and this causes light from the distant Quasar to 'bend'. The nearby galaxy forms a 'gravitational lens' that acts like a convex lens and magnifies the image of the distant Quasar, making it appear much brighter.

What is interesting to us here is that the light output from Quasars varies in a regular pattern and, since they have such huge redshifts they ought to display time dilation on a grand scale, but they don't! M.R.S. Hawkins of the University of Edinburgh, Royal Observatory Scotland, analysed a large sample of Quasars with redshifts as high as three. Without relativity, this implies that they would be travelling at three times the speed of light and so one would certainly expect to find time dilation. He found that "the timescale of quasar variation does not increase with redshift".

That is, there is no time dilation. If there is no time dilation, then there are no relativistic speeds and therefore no expansion of the universe. It is as simple as that.

But it was only the followers of the Big Bang who said, in the first place, that in a stationary universe there should be no time dilation. Can we trust someone with a name like 'Big Banger'?

Does a static universe really not predict time dilation? The answer is yes and no!

No, in that Tired Light and a static universe do not predict time dilation since there are no relativistic effects.

Yes, in that it provides an effect that will give similar experimental results – especially when one considers the uncertainties involved. This Tired Light theory applies the physics of transparent materials to space and so a good place to start is to look for an effect in everyday optics that gives 'time dilation' effects. The place where we find this effect is in fibre optics and is called 'pulse broadening'. White light is made up of different colours and when a pulse of white light is sent down a fibre optic, the different colours of light travel at different speeds. This is known as 'dispersion'. In glass, red light travels the fastest and arrives first whilst blue light travels the slowest and arrives last. What started out as a nice sharp digital pulse of white light ends up as a broad 'blur'. This is pulse broadening and in communication can cause digital pulses to overlap and become confused. Hence, for communications, we use waves of one frequency only.

This is the same situation that we have in space with supernovae. An exploding supernova is a single sharp dramatic event producing light of all frequencies (colours) in a single pulse. This multicolour pulse then travels through the intervening space for millions of years until it is detected here on Earth. With Tired Light, the photons of light making up this pulse are repeatedly absorbed and re-emitted as they travel and, on each interaction with an electron in the plasma of space there is a delay. The total delay in a particular colour arriving depends upon both the delay between being absorbed and re-emitted (known as the 'relaxation time') and upon how many collisions that photon makes on its journey. Both the relaxation time and the number of collisions are different for different colours. Hence different colours will travel at different speeds and so whilst all the colours where emitted at the same time, some will arrive before others. The effect of this is to broaden the supernova pulse; that is, the light curve will be 'dilated'.

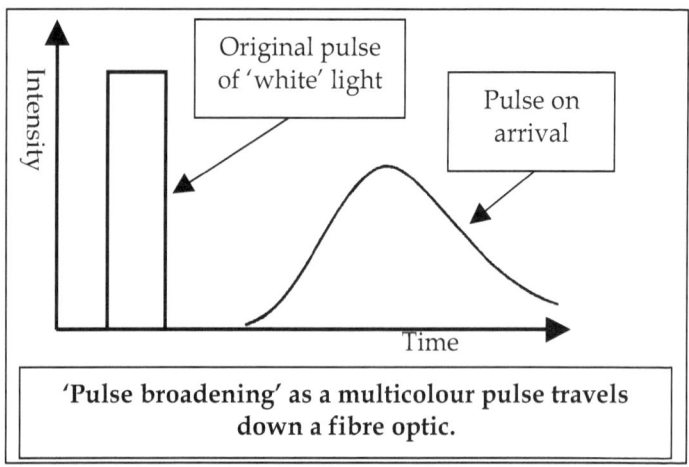

'Pulse broadening' as a multicolour pulse travels down a fibre optic.

In measuring the time dilation, astrophysicists examine the tail of the light curve. The radioactive decay processes cause the composition of the Supernova to change as an isotope of one element decays into another and another until they reach stable iron.

These elements are supposedly formed as regular as clockwork and, once formed, will burst through the photosphere and give out characteristic emission lines. The astrophysicists take the light from the supernova and pass it through a spectroscope. They then divide the spectrum into chunks representing a short range of wavelengths and wait for the characteristic wavelengths from each new element to appear in the spectrum. After that, they then measure the time interval between these characteristic wavelengths appearing. What they claim is that the further away the supernova, the longer it takes for these features to appear in the spectrum. To put it another way, they measure the time interval between different colours of light arriving and thus could well be measuring pulse broadening and not time dilation!

Tired Light tells us that all colours suffer a delay proportional to the redshift, z. This has to be added on to the original time interval between these colours being emitted by the supernova to find the total time interval here on Earth. Consequently, Tired

Light says that supernova multicolour light curves will be dilated and the dilation will increase with the redshift – but does this agree with observation?

In these measurements, the uncertainties are huge and to understand this it is probably best to give an example. In 1997, Riess, Filippenko and Leonard published a scientific paper in the Astronomical journal (reference 1997AJ..114..722R) that was one of the first to claim that a static universe could be discounted. It should be noted that these three are high powered scientists who deserve our respect and it is because of the respect that they deserve that I cite them here. We can trust the results. They examined the spectra of SN1996b which has a redshift of 0.574 over a period of 10.05 days. In an expanding universe, the supernova was expected to age by 6.38 days but with a static universe, with no pulse broadening, the aging should be zero. This is what was expected by the two theories, but what did the results show? Measurements showed an aging of 3.35+/- 3.2 days and the researchers claimed that they were consistent with an expanding universe but not with a static one.

In an expanding universe the dilation should be 6.38 days, in a static universe with no pulse broadening the Big Bangers tell us it should be zero - and then we find that the measured value is 3.35 days. One might be forgiven for thinking that the results showed a draw between the two theories. But the uncertainties are huge at 3.2 days. Uncertainties in science are a science in themselves but basically, they add up the odds of where the result may lie. What the results really show is that the best guess of the time dilation is 3.35 days. However, there is a good chance that it could lie within 3.2 days at either side of this value, so it is likely to be 3.35 days but there is a good chance that the value lies anywhere between 0.15 days and 6.55 days. Since the time dilation of 6.38 days predicted by the expanding universe lies within this range, it is 'consistent' with the results. Also, since the time dilation of zero as supposedly predicted by the static universe lies below 0.15 days and outside the likely range then this result is inconsistent with that theory. However, there is still a chance that the actual value lies outside the quoted range and includes zero. Whilst odds of this happening are small, they are better than those of an outsider in a horse race and they still win. Remember the 1967 Aintree Grand National when Foinavon, the hundred to one outsider, came in first to win? I had money on that, twelve and one half pence each way!

This result and others like it were seized upon as the 'proof' that the universe is expanding! We are often told that supernovae time dilation 'proves' that the universe is expanding and asked to believe in it. Here I have provided a sample from the published results and ask the readers to decide for themselves. Are you convinced? It is true that later measurements were more precise but they still left a lot to be desired. However, with this Tired Light theory we expect a time dilation due to the 'pulse broadening' and so this Tired Light theory is also consistent with these results, so now, there is no decision to be made since 'time dilation' does not discriminate between the two theories as they both predict it!

If you are still not convinced that the Big Bang theory is wrong about time dilation then remember the quasars. If you remember, the light output from quasars also varies periodically, varying over periods of 100 days or so. The only difference is that with quasars, the variation is a continuous event instead of the one off, 'flash in the pan' event of a supernova. Since quasars have extremely large redshifts, we would expect to see relativistic effects in the light curves of these too, and in particular, we would expect to see time dilation. That is, we expect to see that the further away the quasar, the greater the redshift and the longer the time period over which the light curve will vary.

We don't! There is no time dilation at all with quasars even though their redshifts are extreme - equal to and often larger than the supernovae used to 'prove' the expansion of space. Light curves from quasars with redshifts over $z = 3$ have been determined and, in non-relativistic physics, this would imply a 'speed' of three times the velocity of light! We would certainly expect huge time dilations. But there are none. The time period is the same as those with only one tenth of the redshift.

In the Tired Light theory with pulse broadening, we would expect this. 'Pulse broadening' spreads out the light curve since the colours take different times to travel to us. The effect is only apparent in a 'pulse' such as a supernovae explosion. With the continuous light output from a quasar, the effects of pulse broadening are lost since they are only apparent at the beginning and end of the arrival of light. We were not here to see the light 'first' arrive and I certainly won't be here to see the 'last' light arrive from the Quasar. During the lifetime of the quasar, the light curves for each colour will overlap and arrive together. That is, at any instant, there will be photons depicting every colour arriving at the Earth, but they were not all emitted at the same time. The effects of pulse broadening will not be seen in the light from Quasars.

Pulse broadening has another effect on the light curves of the supernovae. Regardless of whether the light curves are 'dilated' or not, we still have the same number of photons given out by the supernovae. The supernovae light curves show the intensity of light plotted against time and so the area underneath these curves represents the number of photons given out by a particular supernova event. Even though the light pulse is broadened, the area under the light curve must remain constant since we still have the same number of photons. With Tired Light's pulse broadening, the more the light curve is 'stretched', the smaller the maximum amplitude must become so that the area remains the same.

In practical terms, this means that distant supernovae look dimmer than they should be. I will repeat that, with Tired Light's pulse broadening, distant supernovae will appear dimmer than they should be!

We have met this before. However, the Big Bangers told us that this effect was separate from time dilation and that it showed that the rate at which the Universe was expanding was increasing ie the expansion of the universe is accelerating. In the Big Bang theory's 'relativistic time dilation', the maximum brightness should remain unchanged. When the Hubble diagram was extended to large redshifts it was found that distant supernovae were about twenty percent dimmer than they should have been when compared to their redshift. But this goes against the Big Bang as the area under the light curves must remain constant with relativistic time dilation. The Big Bangers had to resort to declaring that the expansion of the universe is accelerating with the energy needed for this to be provided by 'vacuum energy'. To explain the dimming they have to rely on a 'vacuum' not being a 'vacuum' at all, but a vacuum that somehow has energy which can be used to drive the extra expansion.

In this Tired Light theory, it is different colours of light travelling at different speeds and thus arriving at different times that causes the dimming. Since the light curve has been broadened, the light intensity 'at maximum' has been reduced and so the supernovae appear to be dimmer.

At the time of publishing this book, a nice little twist to the type 'Ia supernova' story occurred. Supernovae further and further out were found and observed with their redshift and apparent magnitude measured in order to confirm one theory over another. What did they show? Well, actually they showed all the theories to be wrong! Did that mean that the Big Bang theory was finally rejected I hear you ask? Well no, don't be silly, Big Bangers would never do that! What they said was that if the Big Bang theory was correct then there must be something wrong with the supernova! That is, they suggested that the supernova Ia were not all the same. Or to put it another way, don't blame the theory, blame the results!

But what of the Tolman surface brightness test? Didn't that prove the universe to be expanding? This was another test devised to 'prove' expansion. In an expanding universe the apparent brightness of galaxies falls off much more rapidly with distance than it does in a stationary one. Furthermore, by comparing two similar galaxies at different distances away, one can calculate how much dimmer the more distant one should be in both an expanding universe and a static one and these predictions can be compared to observed results. The original data was from relatively nearby galaxies and was said to show that the universe was expanding and not static. However the results were not very convincing and so what was needed was data from much more distant galaxies. The Hubble telescope was used to look at galaxies further and further away (the

Hubble Deep Field HUDF) – so far away that the proof of an expanding universe should be blindingly obvious. Unfortunately for the Big Bangers, when these results were used, they agreed not with an expanding universe but with a static one!

Thus there is no direct evidence that the universe is expanding. Big Bang codsmology relies on 'inflation' and this has yet to be shown to exist. The principle of conservation of energy has to be suspended so that the loss of energy of redshifted photons can be explained. Vacuum energy has to be introduced to explain why distant supernovae are dimmer than expected. The much-vaunted time dilation in supernovae light curves should also appear in the light curves of quasars. It does not, and this shows that relativity is not playing a part. If there is no relativistic time dilation then beyond doubt, even at these huge redshifts, the quasars are stationary. Add to this the fact that for the Big Bang theory to work, around 95% of the universe must be in some 'dark' form and furthermore, a dark form that no one has been so far able to locate! The Big Bang theory is so sick that if it was an animal, the Vet would have put it down long ago!

However, the Tired Light theory, where photons of light lose energy due to interactions between the photons of light and electrons in the plasma of space has many successes. It explains the paradox between the Hubble constant, H and the combination of three very common constants, the Planck constant, and the radius and mass of the electron. Furthermore, using known data for the number of electrons in each cubic metre of space, we can calculate the actual value of the Hubble constant and this agrees with the measured value.

This Tired Light theory predicts an exponential shape to the Hubble diagram. The exponential shape is straight for small distances, which means that local redshifts are proportional to distance as Hubble first found. The exponential graph starts to bend upwards at redshifts above 0.2 and this is known to happen from observation. Furthermore, if one ignores the relativistic 'corrections' applied to supernovae data (that is, forget about space expanding), the data fits the Tired Light, exponential Hubble relationship best of all.

This Tired Light theory provides a realistic, everyday mechanism by which the photons lose energy and become redshifted and it also explains where this energy goes. As a result, the origin of the CMB is explained. In being redshifted by the plasma clouds, photons of light lose energy to the electrons in the clouds and these electrons then radiate this energy as the CMB. Calculations show that this radiation is microwave. The electrons emit the radiation by thermal bremsstrahlung and since the clouds are in thermal equilibrium - whereby the rate at which they gain energy from the redshifted photons is equal to the rate at which they lose it; the radiation emitted is 'Black Body' radiation. We know the temperature of the plasma clouds and this enables us to determine the wavelength at which the intensity of the CMB curve peaks and we find that it agrees with observed measurements of the CMB curve.

This Tired Light theory predicts a time dilation in supernovae light curves, which is known to exist. It also predicts that distant supernovae will appear dimmer than expected and that there will be no time dilation in the continuous light curves from quasars. From observations all this is known to be true in our Universe.

Hence the universe is not expanding.

Chapter 18. A dialogue on the two chief World systems: Tired Light and the Expanding Universe.

The Big Bang theory has become more of a religion than a science. It has become a part of both school and college syllabuses where it must be learned verbatim and reproduced word for word in examinations if the student wishes to pass – instead of being treated exactly as it is, a theory full of holes! Consequently, it is not surprising that people look at one strangely when it is suggested that the whole thing is incorrect. We have been told all along, that there is no doubt about this theory. We have been told time and time again that the Universe started in a Big Bang and is expanding. Who was it who said, repeat a lie often enough and it becomes the truth?

Any scientist who dares to challenge this theory is treated as a leper or at best an eccentric to be ignored. There are many stories about research papers being rejected for no other reason than they produce results that contradict the 'Big Bang' theory. If they manage to get these papers published at all, then it can take years before they come to print. It is as if some sort of censorship exists. Anything that challenges the 'Big Bang' theory either does not get printed at all or has great difficulty in reaching the printed page.

This censorship does not stop there. A popular science magazine – The 'New Scientist', recently published an 'open letter' signed by a large number of scientists who are doubtful about the 'Big Bang' theory. In this 'open letter,' they complained that funding for research is being withheld and only those research projects that are designed to support the 'Big Bang' theory are able to receive the necessary funding that is vital to any project. When one wants to investigate known phenomena that go against the 'Big Bang' theory in detail, then it is said that the necessary funding is withheld.

So, when we said that the expanding universe theory really has become a religion, we were perfectly correct. The Big Bang theory is held in place by bigotry and censorship and is reminiscent of the battles with the church over whether the Earth is at the centre of the Universe or whether it orbits the Sun.

The only difference here is that it is not a case of 'Science versus the Church' but more a case of 'Science versus Science!'

In circa 1600, in an attempt to break the strangle-hold held by the Church, Galileo published a book. In this book he created three characters who would discuss the different theories of whether or not the Earth was at the centre of the Universe.

These characters or 'interlocutors' as Galileo called them (an interlocutor is just someone who takes part in a conversation) included Sagredo, an intelligent businessman who really wished to understand all the arguments and make up his own mind. There is Simplicio, a rather unintelligent man who believed that the Earth was in the centre of the Universe and then there was Salviati, a man who comes over as intelligent and appears to have all the answers and speaks on behalf of Galileo.

These three characters and their discussions led to the Vatican demanding that Galileo come to Rome to explain himself and ultimately, to be punished. Galileo had been under the impression that the Pope had given him permission to write this book. However, this 'permission' was on the understanding that he include within the book, certain points that the Pope felt important. Unfortunately (or, some may say foolishly) Galileo put these ideas into the mouth of Simplicio, the least intelligent of the three and, as may be understandable, the Pope was not too happy about this. Additionally, this was not a Pope who one would want to upset, he being the one who had insisted that the birds in the Vatican gardens be killed - on the grounds that their 'twittering' prevented him from thinking Holy thoughts. In the end, Galileo was led to believe that he was going to be tortured (but he wasn't), made to buy back all the books and put under house arrest for the rest of his life.

Since we appear to be in a similar situation here, it might be a good idea to reincarnate these three men in order to debate the two theories of 'Tired Light' and the 'Big Bang'. I trust that I will not meet the same fate as Galileo did!

Dialogue On The Two Chief Cosmological Models

The first Day
Interlocutors
Salviati, Sagredo and Simplicio

It occurred that a number of casual conversations had already taken place at various times by these gentlemen but rather than satisfying their quest for knowledge it had actually whetted their thirst for more. As a result they prudently agreed to allocate certain days on which all other business would be set aside to allow them to ponder upon the magnificent wonders of God in the heavens. They decided to meet in the palace owned by the eminent Sagredo and, after the usual and short exchanges of salutations, Sagredo began as follows:

Sagredo. "Experiment tells us that light travelling to us from distant galaxies is redshifted. Do either of you disagree with that?"

"No," said Simplicio. "We both agree that the light is redshifted as it travels through space, that is, it has a longer wavelength on arrival than when it set off. It is in how we explain this increase in wavelength in which we differ. In the Big Bang theory we believe that the Universe started in a tiny point or 'singularity' and has been expanding outwards ever since. We explain the increase in wavelength as being caused by space itself expanding and 'stretching' the photons of light as they travel towards us. This increases their wavelength. A photon travelling twice as far, has its wavelength stretched by twice as much and this gives the Hubble relationship between distance and redshift."

"Certainly," said Salviati. "The light does have a longer wavelength when it arrives, than when that same light set off. We would not disagree with experimental results. In 'Tired Light,' we believe that, apart from the individual motion of the heavenly bodies, overall, the Universe is static, not expanding and, perhaps infinite. Photons of light have a longer wavelength on arrival than when they set off. This means that the light has a lower frequency and therefore less energy on arrival than when that same light set off. We say that, on its way through intergalactic space, the photons bump into electrons and lose a little bit of energy each time. This reduces the total energy of the photon of light, reduces its frequency and thereby increases its wavelength. We explain the Hubble relationship, as a photon travelling twice as far will make twice as many collisions, lose twice as much energy and therefore experience twice the redshift."

Sagredo asked, "But for some galaxies, the light takes hundreds of millions of years to reach us, so in the Big Bang theory, are you saying that in all that time the photons of light do not collide with anything at all?"

"In the 'Big Bang' Theory, that would be the case" replied Simplicio. "Of course dust in space will absorb some of the light and make the galaxy appear dimmer but the photons arriving at the Earth have travelled for up to a few hundred million years without colliding with anything."

"This is one of the many things that we find hard to believe," said Salviati. "It is known that there are between 0.1 and 10 electrons in each cubic metre of space; therefore it is difficult for us to say that photons of light can travel for hundreds of millions of years and yet not hit several of these electrons. Further more, we not only know how many electrons there are out there in space but we can use our knowledge of Physics to calculate how often this will happen. It turns out that a photon of visible light will make one collision, on average, every seventy thousand light years or so. We calculate this using the Physics that everyone agrees to be correct. Even if one believes in the Big Bang theory then one must take this effect into account and deduct the redshift caused by this continual absorption and re-emission from that which is measured by experiment. It turns out that when one calculates the Hubble constant due to Tired Light, one finds that the predicted value agrees with the experimental one. There is nothing left over for 'expansion' effects. That is, we don't need 'expansion' or a 'Big Bang'. Using everyday, accepted, Physics, one can predict the value of the Hubble constant and the predicted value is in agreement with the measured value. This is the way Physicists should work. Theory should agree with experiment."

"Well," said Sagredo. "How does the 'Big Bang's predicted value of the Hubble constant agree with measurement?"

"The Big Bang theory cannot predict a value for the Hubble constant," said Simplicio. "We know that the light is redshifted and we put this down to an expansion of space. We measure the Hubble constant and this tells us the rate at which the Universe is expanding. We do not predict a value for the Hubble constant, H and compare that value to experiment; we say that there is an expansion and that the observations tell us what that value is."

"So there is no way of using this experimental result to check if the Big Bang theory is correct?" said Sagredo.

"No," replied Simplicio. "The redshift tells us that the Universe is expanding and the measured value of the Hubble constant tells us the rate at which it is expanding."

Sagredo continued, "But with Tired Light, one can calculate the predicted value of the Hubble constant and then compare that to the value from experiment and we find that they agree?"

"True," said Salviati. "This is one of the reasons we believe the 'Tired Light' theory to be correct and the Big Bang theory wrong."

"But it does not end there," said Salviati. "We find that the measured value of this 'rate of expansion', the Hubble constant, is equal to a combination of three very common constants associated with light interactions. These constants are the Planck constant (h), the classical radius of the electron (r) and the mass of the electron (m). The measured value of the rate of expansion, H is very similar to (hr/m) for the electron in each cubic metre of space. Now, I know that objects themselves do not expand in the Big Bang theory because of forces between the constituent particles, but if we are to believe in the Big Bang theory for just one moment, then we must also believe that the space occupied by a ruler one metre long expands by an amount equal to the combination of these three constants each and every second and put this down to pure chance. That is, if one picks up a one metre ruler from the laboratory, then the 'Big Bang' theory tells us that the space occupied by the ruler increases in length by an amount equal in size to 'hr/m' metres each and every second. Now, this is nonsense because what has the length of a metre ruler got to do with the electron? Nothing at all! Furthermore, In the Big Bang theory, the age of the Universe is related to the inverse of the Hubble constant and so that too must be related to a combination of these same three constants (m/hr). This is what the Big Bang theory is asking us to believe and I for one refuse to do that. I ask you, what has the age of the Universe got to do with the electron?"

"This is just numerology" replied Simplicio. "There are lots of cases where a scientific quantity can be expressed in terms of other things. For instance, the temperature of the Cosmic Microwave Background Radiation is 2.7K and 2.7 is just the sine of sixty degrees multiplied by the square root of ten. That does not mean that they are related, it is just a coincidence of numerology."

"It is not as simple as that," said Salviati. "We must take 'dimensions' or the units into account too. The sine of sixty and the square root of ten have no units and although the product of the two may well give the same value as the CMB temperature they are not a 'temperature'. The value of the Hubble constant in the strange units astrophysicists use is about 64 km/s per Mpc. In the usual units of metres, seconds etc, it has the value of 2.1×10^{-18} 'per second'. This is the same as 'hr/m in each cubic metre of space' or 2.1×10^{-18} 'per second'. Not only do they have the same value but they also have the same units. Or, to put it another way, whilst these two numbers are normally expressed in different units, they are effectively measuring the same thing. Once we put them both in the same units say, 'elephants' then the numbers in 'elephants' coincide! Furthermore, scientists do not like coincidences. We smell a rat whenever a coincidence happens and we only discount the phenomena as a coincidence when we have made absolutely sure that there is not some sort of relationship. Look at Kirchoff and Weber. Both measured the speed of an electrical current down a wire and found it to be the speed of light. They put this down to a coincidence when it most certainly was not and missed a much greater discovery, that of the electromagnetic nature of light. In any case, with the 'Tired Light' theory we would expect a 'coincidence' of these sorts because the Tired Light theory predicts that the two <u>are</u> related.

We start from first principles, using a known physical interaction between light and matter (that is, electrons) and calculate the value of H. The value that we arrive at is 2n(hr/m). We know that n has a value somewhere between 0.1 and 10 and experimental results tell us that it is 0.5. The size of the Hubble constant being equal to hr/m is not an embarrassment to the 'Tired Light' theory; it just means that there is about one electron in every two cubic metre of space and we expected something like that."

"But what about the CMB?" Simplicio asked. "I thought this actually proved that the Big Bang theory was correct?"

"When it was found that the Universe was expanding," said Simplicio. "Two theories were put forward. One was the 'Big Bang' theory itself and the other was the 'steady state' theory. The Steady State theory said that as the Universe expanded, new matter was created in the spaces between so that overall, the Universe stayed the same. In the 'steady state' theory, there should not have been a CMB. But in the 'Big Bang' theory, once all the matter and anti matter had settled down, leaving us with the matter from which the Universe is now made, an unbelievable amount of photons were left over. As the Universe expanded, the wavelength of these photons stretched and became more and more redshifted until they resembled that given out by a 'black' body'. Gamow calculated the temperature of this black body as it should be now and found it to be 2.7K. When Penzias and Wilson actually found the CMB and found it to correspond to that of a 'black body' at around 5K then this proved the 'Big Bang' theory correct".

"That is not quite true," said Salviati. "The discovery of the CMB only showed that of the two theories, the 'Big Bang' was the better of the two and we must remember, that both of these two theories accepted that the redshift was due to the expansion of the Universe. The discovery of the CMB only showed that the 'Steady State' version of an expanding universe was wrong, it did nothing to show that Tired Light's static universe was wrong. Furthermore, Gamow's value of 2.7K was derived at his first attempt. As he perfected his model the expected temperature increased to 'at least 5K' and finally, just before the discovery of the CMB, the predicted value was as high as 50K. On the discovery of the CMB at 5K all Gamow's 'improved' predictions were conveniently swept under the carpet and Big Bangers reverted to the original prediction of the temperature of 2.7K from twenty years earlier. Thus, conveniently forgetting that the improved theory was now predicting much higher estimates of the CMB temperature than experiment showed, they claimed the discovery as a universal victory over all theories."

"But," said Simplicio. "He did predict the temperature first at 2.7K and that is what it was found to be!"

"Let's not get carried away here Simplicio," said Salviati. "The temperature of the CMB never was a 'prediction' because he already knew of Mckellar's work. If you remember, McKellar had shown in the mid 1940's that there must be a radiation everywhere in the universe at a temperature of around 2.7K. Gamow already knew of this value before he came up with his predicted value of 2.7K and so it is hardly a 'predicted' value.

"Can the 'Big Bang' theory predict what the CMB temperature should be now?" Asked Sagredo.

"Well no," said Simplicio. "The 'Big Bang' theory predicts its existence and we use the experimental value to find out what it is exactly."

"So, again," asked Sagredo. "One cannot use experiment to verify the Big Bang theory?"

"In a way, we can," replied Simplicio. "The 'Big Bang' theory tells us that the CMB should be there and experiments tell us that it is, but the actual value is found from experiment."

"Well," said Sagredo. "What does Tired Light say about CMB?"

"Tired Light," said Salviati. "Tells us that on their journey, the photons of light are continually absorbed and re-emitted by the electrons in the plasma of intergalactic space. Every time an electron absorbs or re-emits a photon, the electron recoils and thus some of the energy of the photon is transferred to the electron. As a result, the wavelength of the photon increases. Now, since the recoiling electron undergoes an acceleration, it re-radiates this energy as a secondary photon by a process known as bremsstrahlung. This secondary radiation is the CMB".

"But can the 'Tired Light' theory find a value for the temperature of the CMB?" Asked Sagredo.

"We can calculate the wavelengths of this secondary radiation," answered Salviati. "We find that it is in the microwave region. We know how much energy is transferred to the electron on each interaction and it is then a simple matter to calculate the wavelength of the photon emitted when this energy is re-emitted. Most CMB photons have a wavelength of 2.1mm and these would be created by an original photon of ultraviolet light being absorbed and re-emitted by an electron".

"But the CMB spectrum obtained is that of pure black body radiation" said Simplicio "It is continuous, peaks in the middle and the intensity of the radiation falls off in two 'tails' as one goes further from the peak value. Only the universe cooling down from the 'Big Bang' could produce this sort of spectrum".

"That is not the case," replied Salviati. "We know that the plasma with which the photons interact are in the form of clouds and we also know the temperature of these clouds. If the CMB were, as the Big Bangers suggest, solely due to the redshifted remnants of the initial fireball, then there should be no relationship between the CMB temperature and the plasma clouds. Let us examine this in more detail.

In Tired Light the photons lose energy to the recoiling electrons as they are absorbed and re-emitted. We can, and do, calculate how much energy is lost to the recoiling electron and find that when this energy is re-radiated as secondary radiation, it is in the microwave region and so forms the CMB.

It is known that the CMB has peak intensity at a wavelength of 2.1mm, that is, photons with this wavelength contribute most energy to the CMB spectrum. Secondary photons with a greater or smaller wavelength than this contribute less and less energy to the CMB spectrum. In Tired Light, secondary radiation with a wavelength of 2.1mm is produced when ultra violet light is absorbed and re-emitted – and there is plenty of UV radiation around. Visible light loses less energy to the recoiling electron and so less energy is contributed to the CMB spectrum by light.

Whilst X-Rays and gamma rays transfer more energy to the recoiling electron, they are few and far in between and so we get fewer secondary photons and a smaller energy contribution to the CMB spectrum. Hence the intensity peak is formed at a wavelength of 2.1mm in the CMB spectrum.

To show that the secondary radiation formed in Tired Light is in the microwave region alone, is enough evidence to cast doubt on the Big Bang interpretation of CMB but it doesn't stop there. Let's have a look at the temperature of the plasma clouds, as here we find yet more evidence that the Big Bang interpretation is incorrect.

We know that the temperature of the plasma clouds in intergalactic space is between 10^5 and 10^6 Kelvin and from this we can determine the average kinetic energy of an electron in these clouds. The point where the energy of the incoming photon is equal to the initial kinetic energy of the electron that it interacts with, is of interest and should mark a watershed in the CMB curve. If the energy of the incoming photon is much less than the kinetic energy of the electron, then the electron's motion will not change drastically but will only be modulated by the effects of photon absorption. If the energy of the incoming photon is much greater than the initial kinetic energy of the electron, the effects of photon absorption will dominate. The point where the photon energy is equal to the electron kinetic energy marks the point where one effect finishes and the other starts and should show up as a watershed in the CMB spectrum. When we do the calculations and determine the wavelengths of these secondary CMB photons given off when photons interact with electrons of the same energy, we find them to be in the range 0.076mm to 7.6mm - and it is in this range of wavelengths where the intensity of the CMB curve should peak. Again, this is consistent with observation as the wavelength at which the intensity of the CMB peaks is found to be 2.1mm. The uncertainty is due to the uncertainty in determining the temperature of the plasma clouds. Here we have introduced a new parameter - the temperature of the plasma clouds, and yet the Tired Light theory is still consistent with observation. However, the relationship between the CMB and the temperature of the plasma clouds does not end there, as there is another interesting twist in our story.

There are other interactions between photons and electrons and one of the more common ones is the Compton Effect. Here, the photon is scattered at an angle to the original direction in which the photon was moving. There is an exchange in energy between the photon and the electron but the photon is scattered. It does not arrive at Earth since the direction of the photon has changed. 'Which way the energy goes' in the Compton Effect depends upon the energy of the photon compared to that of the electron. If the photon has more energy than the electron, then the electron gains energy and the photon loses it. This is what is normally known as 'Compton Scatter'. If the energy of the photon is less than the energy of the electron then the photon gains energy from the electron. The electron slows down and the photon gets a boost in frequency. This is known as 'Inverse Compton Scatter'.

When the energy of the photon is equal to the energy of the electron it collides with, there is no exchange of energy by Compton Scatter. This is the point in the Tired Light theory where virtually the whole of the CMB spectrum is formed! Photons with energies equal to the kinetic energy of the electrons they interact with produce the microwave CMB - and yet this is the very same point at which Compton scatter is no longer having an effect. Is it because the two effects, Compton and Tired Light, are competing for electrons to collide with? Any photon that has suffered Compton Scatter is deflected from its original path and will not arrive here on Earth. Only those photons that have undergone recoil interactions will continue in a straight line and arrive on Earth. At the point where the photon energy is equal to the kinetic energy of the electron in the plasma cloud, there is no exchange of energy between the photon and electron fields by the Compton Effect. Tired Light Effects dominate and this is where the CMB is produced."

"Since it is true that the peak of the CMB curve occurs at wavelengths which can be derived from the temperature of the plasma in intergalactic space," asked Sagredo, "How is this experimental result explained in the Big Bang theory?"

"It is a coincidence," explained Simplicio, "There is no way that the temperature of the plasma can be related to the peak of the CMB curve. The Big Bang theory makes no attempt to explain it, since it is just a fluke of numbers."

Salviati smiled and said," In Tired Light, we not only expect the peak of the CMB curve to correspond with the plasma temperature but we explain why it happens and perform the calculations to show that the theory agrees with experiment. With Tired Light we have the mechanism by which it happens, the reason why it happens, the calculations to find the predicted wavelength and this is in agreement with observation. In science this is known as causality and indicates proof."

"I had heard," said Sagredo. "That the 'clumps' in the CMB and their arrangement gave considerable support to the Big Bang and inflation theories. Is that so?"

"Correct," said Simplicio. "Inflation predicted these clumps and there they are!"

"That may be the case," said Salviati, "but the trouble is that these 'clumps', or at least the larger ones, are arranged symmetrically around our solar system and our own galactic plane. Now if these 'clumps' are the seeds from which planets, stars etcetera were formed, then they must be at the very beginning of our universe and so it seems inconceivable that they are aligned with our galaxy as it is now. In Tired Light, we say that these 'clumps' are just the effect of nearby plasma clouds and so we would expect them to be related to our galaxy."

"How does the Big Bang theory explain the alignment of the larger 'clumps' in the CMB with our galactic plane?"

"Some say it is a form of gravitational lensing, others say it is a coincidence," said Simplicio. "Our galaxy must pass through these 'clumps' sometime in such a way that they are aligned with our galaxy, and that time just happens to be now."

"A very fortunate coincidence," said Sagredo. "A very fortunate coincidence indeed! However, I think I read somewhere that there are many more photons of CMB radiation than there are photons of light, IR, UV etc. Why is this?"

"Because," replied Simplicio. "In the primeval fireball, when particles were annihilating each other to form photons, and photons were recombining to form particles and anti particles, as the Big Bang cooled, all these photons that now form the CMB were left over, since they no longer had the energy to reform particles. This is why there are just so many of them."

"The 'Tired Light' explanation of why there is an excess of CMB photons is much simpler," said Salviati. "Photons of light travelling to us from distant galaxies will be constantly absorbed and re-emitted as they travel along. Each original photon will produce many secondary CMB photons as they travel through intergalactic space. Hence it is not surprising in the least that there are many more CMB photons than any other sort. In fact, a single photon of light with an original wavelength of 5×10^{-7} m will produce around 300,000 CMB photons in undergoing a redshift of just one!"

"But where does the energy go?" asked Sagredo. "Whichever theory one believes in, the wavelength of the photons is increased and so their frequency and energy has reduced on their travels. Where does the energy lost by the photons on their way go?"

"This is where 'Tired Light' triumphs," said Salviati. "In 'Tired Light' as the photons travel along, they are constantly absorbed and re-emitted losing energy and becoming more and more redshifted with each interaction. The energy lost is transferred to the electron, which eventually radiates this energy and this gives us the CMB. So, the question was, 'where has the energy gone'? In 'Tired Light' it has gone to forming the CMB."

"So what about the 'Big Bang' explanation," said Sagredo. "I don't see why the stretching of space should cause the photons to lose energy. Since they clearly do lose energy, where does this energy go to?"

"It is true that redshifted photons have less energy on arrival than they had when these same photons set off. The energy 'lost' has gone to driving the expansion of the Universe. It has gone to Gravitational Potential energy if you like" said Simplicio. "In expanding, the Universe is gaining Gravitational Potential energy and this comes from the energy lost by the Photons as they travel through space".

"But by what mechanism do they lose this energy?" asked Sagredo "And in any case, I thought that Gravity was slowing the Universe down. If the redshifted photon's loss of energy is going to providing the increased potential energy, then this would not happen."

"Some people introduce relativity to explain the energy loss, said Simplicio. "The actual energy that something has, depends upon your reference frame. They say that the reason that we see the energy as being different is because we are moving away from the distant galaxy and that puts us in a different reference frame and so we get a different value for the energy of the photon."

"Come now," said Salviati. "Whilst we do measure the actual energy of something differently depending upon where we are, the rate at which energy is lost is invariant, that is, it is the same for everyone and the photons are certainly losing energy. If we see the photons losing energy in our reference frame, then everyone sees them losing the same energy in their reference frame - regardless of what it is."

Simplicio shuffled his feet and stared at the ground before him hoping for inspiration. "In the Big Bang theory, we must suspend the principle of conservation of energy. That is, it no longer applies when it comes to space - time. Mass, energy and space are all interchangeable and the energy lost by the photon in being redshifted has gone to curving space - time and this, in turn, goes to driving the expansion" said he, and left it at that!

"But I thought that we have yet to find any evidence of space being curved at all – in fact it has been decided that space is flat and that is why inflation was introduced," said Sagredo. "Perhaps we should move on a little from the actual redshift and look at something else. I had heard that the measured redshifts are quantised. That is, one doesn't get any value for the redshift of a galaxy, but the measured redshifts are always a multiple of a certain basic amount. You either get one of it, two of it, three of it - but never 'one and a half' of this basic amount. Doesn't that go against the 'Big Bang' theory?"

Simplicio replied, "There is not enough evidence to show that the measured redshifts are quantised and we believe that this is not the case."

"But one cannot just ignore experimental evidence" said Salviati, "A number of researchers have reported quantised redshifts and these scientists work for well respected scientific institutions. Of course, it is in the interests of the 'Big Bang' theory to discount this evidence since quantised results cannot be explained by an expanding Universe theory and would imply that it is incorrect. With 'Tired Light' we would explain quantised redshifts by saying that the plasma in intergalactic space is not continuous but forms in 'clouds' in the same way that Hydrogen does. Light would only be redshifted on passing through one of these clouds and, on their journey through intergalactic space, photons of light would pass through one cloud, two clouds, three clouds and this would give us a sort of quantised redshifts. Naturally the clouds would not all be exactly the same size, but that would give us the uncertainty in the size of the 'basic' quantised amount and thus the debate as to whether quantisation exists at all."

"So, What about the 'Tolman surface brightness test?" asked Sagredo. "That is often quoted as proof of an expanding universe."

"True," replied Simplicio. "The surface brightness of distant galaxies will reduce with increased distance and expanding/non expanding models of the universe predict different relationships. When these predictions were tested for three galaxies at different distances, the results were consistent with an expanding universe."

"Unfortunately," said Salviati. "The three galaxies were in the same cluster so it was hardly a fair test and in any case the results did not prove the expanding universe model correct. In an attempt to prove the case one way or another the Tolman test was applied to the data from the Hubble Deep Field data. This contains data for galaxies far, far in the distance and so the differences in the predictions of the two models are glaringly obvious and easy to check. When this was done, the results showed the expanding universe to be wrong and the static universe to be correct. Consequently, the Tolman surface brightness test is now proof of a static universe!"

"There are said to be 'three pillars' of the Big Bang theory, redshift, the CMB and the formation of the light elements and their relative abundances," said Sagredo. "Salviati, you have already explained redshift and the CMB, what do you have to say regarding the third pillar – the light elements?"

"In the 1950's, said Salviati. "This was a major breakthrough since 99% of the Universe was then thought to be Hydrogen. The Big Bang theory, in explaining how Hydrogen and Helium were formed and correctly predicting their abundance, was successful since it explained 99% of the then known universe - and this could truly be thought of as a 'pillar'. However, things have changed since then. Nowadays, in order for the theory to work, 95% (or even 99%) of the universe must be in some dark form, which we have yet to detect. The Big Bang's prediction of the light elements fades into insignificance when one realises that it now only predicts how, at best, 5% of the universe was formed! Worse still, the theory cannot correctly predict the abundance of the light elements at the same time! By 'abundance' we mean 'how much of that element there is compared to Hydrogen'. If the theory correctly predicts the abundance of Helium then it gets the abundance of Deuterium and Lithium both wrong. Similarly, if it correctly predicts the abundance of Deuterium then this time it gets the other two abundances wrong! Big Bangers often quote the prediction of the abundance of the light elements as a success - but conveniently forget that it cannot do it all at the same time! However, we now know that stars produce these light elements in similar abundances and so the third and final 'pillar' of the Big Bang has now long gone."

"But don't galaxies have to be more than about fifty million light year away before the expansion effects become noticeable?" asked Salviati.

"That is correct," replied Simplicio, "Galaxies have to be part of the 'Hubble Flow' before the expansion of the Universe becomes apparent. On the one hand, we have space expanding and on the other hand we have gravity causing local galaxies to move together. Nearby, the effects of gravity are greater than the effects of expansion so we do not get this notion of everything moving away from each other as space expands. Over large distances, ie when galaxies are in the Hubble flow, expansion effects are large enough to make gravitational effects insignificant and it is here that the Hubble relation between velocity and distance holds true."

"Well no," replied Salviati. "The universe is not expanding. The reason why galaxies have to be at least a certain distance away is partly because of local gravitational effects and partly because of statistics. 'Tired Light' depends upon statistics, which are the probabilities of photons colliding with electrons in the plasma of intergalactic space. In order to achieve reliable and repeatable results in statistics one needs a large enough sample. In order that the Hubble law be obeyed, the photons of light must have travelled far enough to ensure that they have made a sufficiently large number of collisions that the statistical results hold true. With 'Tired Light,' nearby galaxies don't exhibit cosmological redshift because the photons have not made sufficient collisions for the statistics to be repeatable."

"Salviati says that the Universe is not expanding" said Sagredo. "Simplicio says it is! If as Simplicio says the Universe started in a 'Big Bang' and everything has been whizzing outwards ever since, then how did galaxies, stars and planets form. Surely they require matter to fall together and not whiz apart."

"That would only be a problem if the matter in the Universe was evenly distributed and moving 'outwards' as if from the 'Big Bang' said Simplicio. If the matter were in 'clumps' - regions where the density of matter was a little greater, then local gravitational effects would cause these clumps to coagulate as they continued with their overall expansion. This would form the galaxies, stars and planets. Evidence of this has been found and this 'clumpiness' shows up as slight variations in the CMB which has now been measured."

"So if the CMB is not entirely smooth, does this mean that 'Tired Light' fails this test?" asked Sagredo.

"Certainly not!" replied Salviati. "Do you remember the quantised redshifts and how we said that this implied that the plasma of intergalactic space was not continuous but was in the form of 'clouds' like Hydrogen? Well if this is the case, then we would expect 'clumpiness' in the CMB. Photons of light would only be redshifted within these clouds and since the CMB is produced whenever the photons are redshifted then we expect these plasma clouds to 'glow' with microwave radiation - each cloud will act as a 'clump' in the CMB. Not only this, but remember how the larger clumps in the CMB are aligned with our own galaxy? In astronomy, large means 'near' and there are many hydrogen clouds left over from the formation of our local cluster and there are lots of plasma in those clouds. As a consequence, we expect these clumps to be aligned with our galaxy."

"But are there any exceptions to the Hubble relationship between redshift and distance" asked Sagredo?

"Yes there are several," replied Salviati. "Time and time again young, massive stars have a redshift in excess of what they should have, considering how far away they are. This is known as the K Trumpler effect. People who support the 'Big Bang' theory ignore these results because it shows that something in the stars themselves is affecting the redshift. It appears that redshift is made up of a part distance effect and a part 'intrinsic' effect – a redshift due to something within or around the stars. The 'Big Bang' theory cannot explain this intrinsic effect and this is why this effect is largely ignored. With 'Tired Light' we can explain it, since young massive stars will be gaseous and would have more plasma around them, thus giving a systematic error due to the increased redshift in the light as it escaped through this environment. The total measured redshift would be that due to the increased density of the plasma around the star plus the redshift that we normally get due to the light travelling through space."

"But there are many regions within stars and Space itself where there are plasmas of great densities." said Simplicio. "Surely, if 'Tired Light' were correct then one would expect huge redshifts in these areas?" He felt pleased with himself as he thought that here he had made a good point.

"No," replied Salviati. "This is the interesting point with 'Tired Light'. In order for the electrons to recoil the plasma has to be 'squidgy'. That is, the forces between electrons in the plasma must be strong enough to allow the electrons to oscillate with simple harmonic motion so that they can absorb and re-emit photons of light and yet weak enough to allow the electron to recoil, thereby allowing it to take up some energy from the incoming photon and thus redshift that photon. Once the density becomes too large, the repulsive forces between the electrons become very large and prevent it from recoiling. A photon can still be absorbed and emitted but since the electrons are prevented from recoiling, the emitted photon is identical to the absorbed one - so no redshift takes place. Another way of looking at this is to say that the greater the density of the plasma, the greater the effective mass of the electron becomes and so the amount of recoil is less. If we were to take this to extremes and think of a piece of glass, then here the electrons are bound within the glass. As light travels through the glass it is continually absorbed and remitted by the electrons within the glass but since the effective mass of the electron is now the mass of the whole block of glass, the recoil is negligible. This is why light is not redshifted on passing through a piece glass."

"But if light is continually absorbed and re-emitted as it travels through space, then how does it manage to travel in straight lines?" asked Simplicio, not going to give up too easily.

"The easy answer to this question," replied Salviati, "is that it is the same process that happens in glass as happens in space. Light manages to travel in straight lines in glass so why should the same not be true in space?"

"But what is the hard answer?" asked Simplicio.

"Conservation of linear momentum" replied Salviati. "The recoil of the electron and the absorption and re-emission of the new photon all occur along the same straight line. In this way, we achieve the linear propagation of light."

"Does not time dilation in supernovae light curves finally prove that the universe is expanding?" asked Sagredo.

"We believe that it does," replied Simplicio. "The tail of the supernova light curve is driven by radioactive decay. The rate at which radioisotopes decay is completely independent of everything including temperature. The only known quantity on which radioactive half-lives are known to depend is relativity. When the redshifts of these distant supernovae approaches unity, they have speeds approaching the speed of light and relativistic effects come into play. We not only expect time dilation but it has been measured and found to exist. Tired Light predicts a static universe and so we would not get time dilation. In Tired Light, the rate of decay of the isotopes produced in the supernova explosion would appear to be the same, no matter what the redshift of the supernova was."

"That is an over simplification," said Salviati. "Firstly, the uncertainties in these measurements are very large and can be almost as high as 95%. That is, almost all of the measured time dilation could be due to experimental uncertainty and so it is not absolutely clear that it exists at all. Secondly, we have the Quasars, which are proving to be the thorn in the side of the Big Bang theory. Quasars have huge redshifts and so they should exhibit huge relativistic effects. They exhibit none at all and so, on the basis of time dilation, the universe cannot be expanding. One cannot 'cherry pick' results, if 'time dilation' of light curves really is due to expansion effects then both supernovae and quasars should exhibit it - and they do not."

"This is an interesting point, Simplicio," said Sagredo. "With high redshift supernova we see relativistic effects in the form of time dilation, but in higher redshifted quasars we do not see time dilation. How do you account for that?"

Simplicio replied, "Quasars are unique objects in the universe and as yet, we still do not understand what they are. As for why they do not exhibit time dilation, we will have to wait until we understand them better."

"But we understand that they have high redshifts and we understand that their output varies periodically," replied Salviati. "What else is there to understand? If the time period does not vary with redshift, then there is no time dilation - implying that the universe is not expanding."

"Yes," said Sagredo. "I understand that, but I thought time dilation specifically disproved Tired Light. How can a static universe create time dilation?"

Salviati said, "Tired Light and a static universe do not produce time dilation effects, they merely create the conditions where we detect similar phenomena. The crucial factor here is that the light from quasars is continuous whilst that from a supernova is a brief pulse of light and thus we need a process that will discriminate between the two cases. That is, brief pulses of light demonstrate effects that could be interpreted as time dilation whilst continuous sources do not. This is not a new phenomenon and actually follows on from Tired Light theory, which is just the theory of how light travels through a transparent medium applied to space. If one sends a pulse of 'white light' down a fibre optic one detects 'pulse broadening'. Different colours of light travel at different speeds down a fibre optic. Red, which travels the fastest in glass, arrives at the other end first whilst blue, which travels the slowest in glass, arrives last. This gives the appearance of the light curve of the pulse having been stretched.

In space, different colours of light will also travel at different speeds due to the number of collisions the photons make with electrons and the delay suffered at each collision, and so the 'pulse' of light emitted by a supernova will also be broadened or 'dilated'. Tired Light says that the further away the supernova, the more collisions are made by the photons on their way and thus we see both a greater redshift and a greater 'pulse broadening'. The further away the supernova, the more the light curve is stretched.

However, for a pulse that was originally very broad, we would only notice these effects at the start and finish of the pulse's arrival. This is what happens with the continuous and variable light output from quasars. We were not here when the first light arrived from a quasar, nor will we be around when the last light arrives at the death quasar. We only see the middle of the 'pulse' when all the photons overlap - but there will be no pulse broadening or time dilation.

Of course this 'pulse broadening' of supernova light curves has another advantage. The area under the light curve must remain the same, as this is a measure of the total number of photons. If the light curve broadens, then the height must reduce so that the area remains the same. In practical terms this means that distant supernova will appear dimmer than they would otherwise appear and this is what is found in our observations of distant supernovae. The Big Bang theory interprets this as being due to the expansion of the universe accelerating due to 'vacuum energy'. Tired Light does not need this abstract term."

"But I thought that it had been proven that the rate at which the Universe is expanding really is accelerating?" asked Sagredo. "Is that not the case, Simplicio?"

"It certainly is!" agreed Simplicio. "In astronomy 'near' means 'young' and 'distant' means 'old'. The Hubble constant found from near and 'young' supernovae is larger than the Hubble constant found from 'distant' and therefore 'old' supernovae. This means that the Universe is expanding at a greater rate now than it did before, that is, the rate of expansion is accelerating."

"Well," asked Sagredo. "Does this not disprove your Tired Light, Salviati?"

"Certainly not!" said Salviati. "You see, when the 'raw data' has been recorded it has then to be 'adjusted' to take account of the effects of dust and so on. However, the Big Bangers don't stop there, as they assume that the Universe is expanding and then 'adjust' the data to take account of the relativistic effects on these supernovae, as predicted by expansion. Having done this, they then find that the distance as calculated from the redshift does not agree with the distance as found by intrinsic brightness! Instead of just saying that the Big Bang theory is wrong, they trump up 'vacuum energy' and 'acceleration' to account for the discrepancy."

"Well, how does Tired Light explain this effect?" asked Sagredo.

"Very simply," replied Salviati. "One takes the data and adjusts it for the effects of dust absorbing some of the light as is the usual practice. However, one stops there. We say that the Universe is stationary and not expanding and there is no need to take account of any relativistic effects. The Tired Light theory derives an exponential relationship between redshift and distance and we find that the data fits the exponential relationship perfectly. There is no need to invent anything - the data proves that Tired Light is correct since the data agrees with the theory".

"But this data is collected from satellites," said Simplicio. "The satellites are run by NASA. Are you saying that NASA is wrong?"

"No," replied Salviati. "NASA put the satellites into orbit and collects the data. It is left to astrophysicists to interpret the data and after a while the data is put on 'open access' where anyone can download and use it. However," continued Salviati, "NASA have been known to make mistakes in the past, and I do not mean technical failures either. For instance, there was the Hubble Space Telescope itself. When the telescope was first put into orbit, it was found that the images were out of focus since allegedly, a mirror had not been ground to the correct shape. One might have expected a fault like this to be detected before the telescope was put into space but it wasn't and astronauts had to do a space walk and fit it with a 'monocle' to improve the telescopes vision. Then of course there was the one hundred and forty million pound Genesis space probe, which had spent over three years in space collecting space dust. In order that the samples would not be contaminated by the Earth itself on landing, a Hollywood spectacular was arranged whereby two stunt pilots would fly helicopters and 'catch' the probe in mid air before it hit the ground. After a prolonged and intensive build up, the World's television news crews arrived at the Utah desert, to transmit pictures of the probe's spectacular descent and 'catch' to an expectant audience of hundreds of millions. What they saw was the probe thundering down to the ground followed by an expensive 'thud' as the probe partially buried itself into the Utah desert! The problem was that the parachutes, designed to slow down the descent, had failed to open. Apparently the reason for this fault, was that someone had installed the switches controlling the parachutes upside down!"

"But mistakes happen," Said Simplicio. "Not everyone is perfect".

"True," Said Sagredo. "But why was it that no one had noticed that the measured value of the Hubble constant was exactly equal to hr/m in each cubic metre of space until 'Ashmore's paradox' came along?"

"Well," said Simplicio, "Astrophysicists do not work in the standard system of units of metres, kilograms and so on, they use astronomical units such as Megaparsecs and so the relationship was not as obvious."

"It must be said," replied Salviati, "That this is not the first time that this has led to mistakes. Take the one hundred and twenty five million dollar 'Mars Climate Orbiter' satellite. It spent a whole eleven months travelling to Mars only to splat straight into the Martian surface. The reported cause of the fiasco was that one group of scientists on the project had been working in pounds and feet, whilst another group of scientist working on the project had been working in kilograms and centimetres. The computer software just could not cope and so the orbiter crashed on arrival."

"So," said Sagredo. "It is clear that we can not rely on an expanding Universe interpretation on reputations alone. We must look at the evidence and make up our own minds."

"But," said Simplicio. "If Tired Light followers such as Salviati discount the Big Bang theory of the Universe, then what are they going to put in its place. In Science, one cannot throw one theory away until one replaces it with another!"

"Firstly", replied Salviati, "To say that we must keep the Big Bang theory and an expanding Universe until we can find a replacement is simply untrue. Let us say that the police were to arrest you tomorrow for a murder that you did not commit. Irrefutable evidence was then found that showed that you did not commit the crime. As a scientist, would you say that you had to remain in gaol until a new and more likely suspect was found? No, you would not. You would insist on being released and the police would go out with renewed vigour to find the real culprit. We have evidence that the Big Bang and the expanding Universe is wrong so let us reject it and then go out and find the truth of how the Universe really is. If anything, clinging on to an old and disproven theory is making us complacent and holding us back."

"Agreed," said Sagredo, "But you must have some thoughts on the matter of what a non-expanding universe looks like."

"Thoughts, yes," replied Salviati. "Such as why should the Universe not be infinite? Time and time again man has considered himself as 'special' and been wrong every time. Why should there be just this 'bubble' in the place where we are? Why cannot the Universe be like it is locally and just go on forever with more and more of the same? "

"For one thing", replied Simplicio, "We have shown mathematically that a Universe must be either expanding or contracting and so it would be impossible to have a static Universe."

"If we took the Earth and Sun, mathematically speaking they have to either collapse or move apart but their kinetic energy keeps them apart. The Earth is in fact falling towards the Sun giving us in effect, a collapsing Earth – Sun system. But, as the Earth falls its speed keeps it all apart. Why can our infinite Universe not be the same? That is, as gravity forces it to collapse, the kinetic energy of the galaxies means that they move past each other and thus the system remains intact."

"But what about Olber's Paradox," asked Simplicio? "In an infinite Universe, wherever we looked there would be a star at the end of our line of sight and so the sky would be light at night. Because the sky is dark, the Universe cannot go on forever."

As light travels through great distances, it is redshifted," replied Salviati. "If this was the case then the 'brightness' of the night sky would be in the longer wavelengths such as microwaves and thus not be visible. A more likely explanation is that opaque dust particles in the intergalactic space get in the way of distant stars absorbing the light and blocking the stars from our view."

"But wouldn't that mean that the dust would heat up and 'glow' thus giving us a bright night sky anyway?" asked Simplicio.

"No," said Salviati. "This is incorrect. An infinite Universe is just a lot of small Universes side by side. If each of these 'small Universes' doesn't heat up then why should the whole?"

"Could you tell us what an infinite Universe would look like," asked Sagredo? "That is, just what is infinity?"

"A discussion on infinity would go on forever!" joked Salviati. "The usual interpretation is a universe that looks the same no matter where one stands within it. In this model, the Universe goes on forever with everything the same, galaxy after galaxy, repetition after repetition".

"It sounds like you have a different version, Salviati. Could you tell us what that is?" asked Sagredo.

"As well as being infinite in distance, it could also be infinite both 'in and out' so to speak," continued Salviati. "Have you seen the Russian dolls where one has one doll inside another inside another and so on? Some have over twenty dolls - one inside the other."

"Yes," replied Sagredo, "They are called 'Matrushka' dolls."

"Well", said Salviati, "We look upwards and see planets, star systems and galaxies. Could that be a part of a large-scale structure? We look down at the atoms and see electrons whizzing around protons and neutrons. We look inside protons and neutrons and see that they are made up of an even smaller structure of quarks. Could our infinite Universe be a 'Matrushka doll' Universe with one Universe inside another, inside another? Then there is time. Is an 'infinite Universe', infinite in time as well as space? Is time just a series of Universes side by side and time is just ourselves moving from one Universe to the next - like moving around a game board at the throw of a die? The possibilities of an infinite Universe are endless!"

"Interesting Salviati," said Sagredo. "But this is just speculation. Aren't these ideas a little far fetched?"

"No more 'far fetched than a Universe that started out of nothing in a Big Bang and has been expanding ever since!" replied Salviati.

"Let's return to the theories themselves" said Sagredo. "We have two theories here. The Big Bang theory, which says that the Universe is expanding and the Tired Light theory, which says that it is not. Is there any common ground between them? That is, could they possibly overlap?"

"No," said Simplicio. "The two theories are mutually exclusive. Either the Universe is expanding or not."

"Well actually," said Salviati. "This is not true. The Tired Light theory just says that the Universe is not expanding now. It does not tell us one way or the other whether it has ever expanded in the past. The two theories could agree in that the universe started in a Big Bang, and expanded as per the Big Bang theory. However, the conditions were such that gravity was just strong enough to arrest the motion and bring it all to a stop but not strong enough to cause it all to collapse again. The Universe would then remain in an unstable, but static, state. In fact, this is one of the possible outcomes of the Big Bang theory itself."

"But doesn't this seem a very unlikely situation" said Sagredo, "I mean, for everything to be 'just so' that the universe expands and the force of gravity has exactly the correct strength to arrest the motion but not cause it all to collapse once more."

"The truth of the matter is that this really is the case," said Salviati. "Even if one believes in the Big Bang, experimental evidence tells us that the density of the Universe is very close to the critical value that would cause this situation to occur. The critical density is the density of the Universe that would enable gravity to bring the expansion to a stop, but no more. Scientists have researched the density of the Universe and found it to lie between 0.8 and 1.2 times the critical density."

"That appears to be an amazing result," said Sagredo. "The density of the Universe could have had any value at all so why is it that the density is so close to the critical density that would bring it to rest and stay at that point forever?"

"It is true that it was slightly embarrassing to the Big Bang theory at first," said Simplicio. "Some said that it was so close that the density of the Universe must actually be equal to the critical density. However the theory of 'Inflation' explained how it came to be this way."

"Yes", said Salviati, "The Inflation theory also explained the variations, or clumps, in the CMB and then it was found that the larger of the clumps were aligned with our Solar System and galactic plane - thus showing that the clumps were local and the theory incorrect."

"Well, your turn", said Sagredo, "How could the Big Bang theory and Tired Light theory agree?"

"We saw before that whilst the Hubble constant could theoretically have any value, if we want life as we know it then this is not the case. The Hubble constant or rate of expansion is governed by gravity. If gravity were to be a little bit stronger than it is at present then the Universe would expand, gravity would arrest the motion and cause it to collapse and 'crunch' in a short time scale. There would not have been time for galaxies, stars, planets and life to form. If gravity were to be a little bit weaker than it is at present, then the 'clumps' would not have coalesced and no structure would have formed in the Universe. The 'dust' would have gone on forever and nothing would form in the Universe – no galaxies, no stars, no planets and no life at all. No, there is only a narrow range of values that gravity and the Hubble constant can have if 'we' are to exist. This is why it is just too amazing to be true that this particular value of the Hubble constant, the one that allows life to form, just happens to be equal to a combination of the parameters of the electron in each cubic metre of space. So, how about if the Big Bang occurs as the books tell us, the light elements form and so on and the Universe expands. But, the density of the Universe and the force of gravity are 'just right' so that it brings the Universe to a halt, but they are not strong enough to bring it all back crashing down. Redshift and CMB would be explained as due to interactions between photons and electrons in a static, non expanding Universe as set out in this Tired Light theory. The Universe would not be infinite but we could live with that."

"Yes," said Sagredo. "But we still have this unlikely scenario where the Universe neither continues to expand or contract."

"Well, this is where we invoke Darwin's theory of evolution and the 'survival of the fittest'. Let's say that there have been many Big Bangs, one after the other. For some of these, gravity was too strong and they collapsed before life could form. For others, gravity was too weak and these universes continued expanding forever and we see them no more. Our Universe though had just the correct conditions for the Universe to expand and stay there. This gives us time for galaxies to form and life to begin. This situation is then more likely as it stays there forever. In this scenario the two theories would agree."

"If this were the case," said Simplicio. "There should still be 'Bangs' going off at present and forming new Universes."

"Maybe there are," replied Salviati. "We have only been observing the distant Universe for the last hundred years or so, and we only have records of local observations for the last few thousand years. This time span is nothing compared to the age of the Universe."

"But when we look deep into space," said Simplicio. "We are seeing back in time and are presented with the history of the Universe before our eyes. Surely we would see the odd 'crunch' of a universe if not a Big Bang."

"But we do see crunches" said Sagredo, deciding that he too would join the speculation, "We see the effects of matter falling into Black Holes and the density increasing and increasing. Could these become big enough to explode, reducing everything to quarks and the whole process start over again?"

"Let's not get carried away here," cautioned Salviati. "As far as present Physics is concerned, this will not happen – but you never know, 'impossible' is a word scientists try to avoid."

"True," sighed Sagredo, "Perhaps we should return to our original discussion and each in turn, go over the reasons for their beliefs. Your turn first, Simplicio."

"I think that the Big Bang theory and the expanding Universe are correct because of four things. Firstly we have the redshift itself. Photons of light from distant galaxies are stretched thus showing that space is expanding. Secondly, we have the CMB, this is the echo of the Big Bang. The theory predicts that there will be a left over radiation in the microwave region and we find that this is so. Thirdly, we can predict the abundance of the light elements and this too agrees with experiment. Fourthly, we have a theory that explains how it all started and where we are all from. Nothing else comes close to this and so I believe that the Universe is expanding."

"We need to look at this in two parts," said Salviati. "Firstly, why the expanding Universe theory is wrong and then what successes the Tired Light theory has. I find it very hard indeed to believe in the Big Bang theory as whilst they claim that redshift is caused by photons of light being stretched as the Universe expands, the theory cannot provide a convincing argument to explain where the resulting loss of energy goes to or even how it disappears. There are too many instances that disagree with the theory – such as the Quasar found at the beginning of 2005, which, according to the expanding Universe theory has a redshift that puts it several billion light years away whilst it is actually contained within a galaxy that is only three hundred million light years away. In the same way, the 'clumps' in the Cosmic Microwave Background radiation are supposed to be (in the Big Bang theory) at the beginning of time and are the seeds from which galaxies and so on formed and we now know these to be aligned with our own galaxy and solar system. This means that the clumps are local. If the clumps are local then the CMB is local – meaning that they are not leftovers from the Big Bang and are certainly not the seeds from which galaxies are made. As one looks deeper and deeper into space, we go back in time and so we should see the Universe developing before our eyes. That is, it should be like looking through an old family photograph album, starting at the last page and gradually flicking through the pages in reverse order until we end up on the first page. On the last page of the album, we see your father as he is now, grey and wrinkly and, as we work our way towards the front of the book, we see him younger and younger, dark haired and handsome, then with his young children, then as he was at his marriage and finally, on the front page, as he was as a baby – we should see his entire life before our eyes, going backwards through his life. When we look deep into the Universe, we should see the same thing but with the galaxies. Nearby galaxies should be older, like our own, well formed and spiral; further away they should be perhaps younger, perhaps be gassy and irregular, then dust and so on until we get back right to the beginning of the universe itself. Iron was formed by supernovae and thus appears later in the life of the Universe and so there should be a 'cut off' distance beyond which there is no iron. What we find is that distant galaxies look just like our own – there is no progression or aging process with distance. In fact a huge conglomerate of galaxies, all perfectly formed, has been discovered with a redshift that indicates that it was formed almost at the beginning of the Universe itself. In the expanding Universe this should not be there, as there was not time for it to form. There is also iron out there deep in space where iron should not be, as there had not been time since the Universe began for the stars to form, live and then die in a supernovae explosion. Then on top of all this we have Ashmore's Paradox, where the measured value of the rate of expansion is exactly equal to a combination of the parameters of the electron. For this to be true, the whole thing is absurd. The experimental results are correct so it must be the theory that is wrong. If this is not bad enough, we find that the wavelength at which the CMB curve peaks can be determined from the temperature of the plasma in intergalactic space. Combine this with the evidence that the ' larger clumps' in the CMB are aligned with our galactic plane and one knows that the Big Bang theory is wrong.

Why do I believe in Tired Light? Well this theory starts with accepted Physics and derives a formula for the Hubble law. It gives $z = \exp(Hd/c) -1$ and this relation has been shown to agree with redshift – distance data from distant supernova. This theory predicts that $H = 2nhr/m$ for the electron and when we insert accepted values for the terms, we end up with a value for the Hubble constant H, which agrees with experimental results. When we work out the magnitude of hr/m, we find it is equal to quoted values of the Hubble constant – something that had not been seen before. This theory predicts a mechanism by which the CMB can be generated and when we substitute accepted values into this formula, we get wavelengths in the microwave region. In this theory, comparing the photon energy with the average energy of electrons in the plasma of intergalactic space predicts a watershed in the CMB curve and when we insert the numbers, we find that the wavelength at which the CMB curve peaks corresponds to the photons of these energies. That is why I believe in Tired Light."

THE END
(of the expanding Universe!)

Acknowledgments

Where to begin?

It takes a great many people to produce a one-man theory and one-man book! Lets start with the book.

There's my wife Valentina who not only drew the 'people' pictures at the beginning of the book but also didn't complain too much when I slunk off to the computer night after night to 'write some more words'. Then there's my Brenda who gracefully went through the whole book checking for typos, grammar and generally completing the first 'edit'. Then there's my good friend Jay who not only put up with my endless ravings (over several pints I might add) as I developed the theory but they also read the first draft and suggested ways in which the book could be improved further.

A big thank you must go to the Editorial Board of "Galilean Electrodynamics" - the peer reviewed scientific Journal that accepted the theory for publication and gave me the much needed boost and confirmed that I was not totally off my rocker in saying that the universe is not expanding.

In order to try out and perfect some of the explanations included in the book I took part in several forums on the Internet and so thanks too must go to these forums for allowing me and others to participate. In particular I would like to thank 'Sci.astro', 'Bad Astronomy' (now BAUT), 'EvC', 'The website of Halton Arp, ' 'Focus' and of course my own 'Big Bang Blasted'. Many people took a great deal of time and patience in discussing the theory and, in doing so, pushed me to extend the theory further than I originally intended. There are too many 'posters' to mention everyone, but I feel I must mention in particular some who I am especially grateful to; Chiropractor 24, Papageno, Sylas, 'Eta C', all of who replied in detail, in depth and with questions that were always particularly constructive.

Lastly, I would like to thank you - since in buying the book, you show an interest and an enquiring mind that hopefully will enable enough of us to stand up and say 'The universe is not expanding!'

Cheers,

Lyndon.

Lyndon Ashmore.

Bringing cosmology back down to Earth!

Bibliography.

"Einstein's Daughter – The Search for Lieserl" Michele Zackheim
"The Extravagant Universe" Robert P. Kirshner. Princeton University Press 2002.
Einstein's Brain - The Lancet (vol. 353, pages 2149-2153) on June 19, 1999
Acosta, H. M. "Eight Factors Effecting Focal Distance and the Moon Illusion." Doctoral Dissertation, Las Cruces, NM: New Mexico State University.
Andrade, E.N.da C. 1950 "Wilkins Lecture; Robert Hooke." Proceedings of the Royal Society. 201A, 439-473.
Annals of Harvard College Observatory. Bibcode: 1895AnHar....34....1B
Arp. H. "Seeing Red." Apeiron. Montreal.
AAVSO "Variable stars."
http://www.aavso.org/vstar/vsots/1298.shtml
Baird, J. C. and Wagner, M. "The Moon Illusion : How high is the sky?" Journal of Experimenetal Psychology, 111, 296-303.
Biographies: University of St Andrews, Scotland. http://www-history.mcs.st-andrews.ac.uk/Mathematics/Hooke.html
Blumenberg, Hans. 1987. "The Genesis of the Copernican World." Cambridge, MA; Massachusetts Institute of Technology Press.
Boorstin, D. 1983. "The Discoverers." New York; random House.
Chance, W.H.S. 1937. "The Optical Glassworks at Benediktbeuern." Proceedings of the Physical Society (London) 49, part 5, N^0 275. 433-443
Christianson. G.E. "Edwin Hubble. Mariner of the Nebulae." University of Chicago Press. Chicago.
Church. J.A. 1963. "Optical Designs of Some Famous Refractors." Sky & Telescope 63, 302-308
Clark. D.H. and Stephenson. F.R. 1977. "The Historical Supernova." New York. Pergamon Press.
Cklerke. A. 1985. "The Herschels and Modern Astronomy." New York. McMillan and Company.

Clark. S. "Towards The Edge of the Universe." John Wiley & Sons. Chichester.
Crowe. M.J. Ed.1998. "A Calendar of the Correspondence of Sir John Herschel." Cambridge University Press.
Crowe. M.J. "Modern Theories of the universe. From Herschel to Hubble." Dover. New York.
Cunningham. P. "Hand book of London." 1850.
Dewhirst. D.W. 1955. "Observatories and Instrument Makers in the 18th Century." Vistas in Astronomy 1, 139-143.
"Dudley Observatory – the History."
http://www.dudleyobservatory.org/History/history.htm
Dyson. F.W. 1915. "Measurement of the Distances of Stars." Observatory 38.
'Espinasse, M. 1956. "Robert Hooke." London. William Heinmann Ltd.
Ferguson. K. 1999. "Measuring the Universe: Our Historic Quest to Chart the Horizons of Space and Time." New York. Walker and Company.
Ferdinand Magellan. The Mariners Museum. Newport Virginia.http://www.mariner.org/age/Magellan.html
Ferdinand Magellan. The Catholic Encyclopaedia.
http://www.newadvent.org/cathen/09526b.htm
Feynmann. R.P. "Quantum Electrodynamics." Perseus Books. Reading. Mass.
French. A.P. "Special Relativity." Nelson. London.
Galilei Galileo. 1964. "Dialogue on the Great World systems." Salisbury translation. Chicago. University of Chicago Press.
Gilman. C. "John Goodricke and His Variable Stars." Sky and Telescope. Nov. 1978
Gould. W.L. 1989. "Small Scale Telescope/ Joseph Fraunhoffer 1786-1826. Sky & Telescope 77, 250-251.
Gribbin. J. "In Search of the Big Bang." Corgi Books London.
Heath. T.L. 1966. "Aristarchus of Samus: the Ancient Copernicus." Oxford. Clarendon Press.
Heath. T.L. 1991 "Greek Astronomy." New York. Dover Publications.
Herrman. D. 1984. The History of Astronomy from Herschel to Hertzsprung." Cambridge University Press.

Herrman. D. "Women Astronomers." Popular Astronomy.
Hirshfield. A.W. 2001. "Parallax." New York. W. H. Freeman & Co.
Hoyle, Burbridge, Narlikar. "A Different Approach to Cosmology." Cambridge Uni. Press. Cambridge.
Jackson. J. 1922. "Early Estimations of Stellar Distances." Observatory 45. 341-352.
Jones. J & Boyd. L. "The Harvard College Observatory: The First Four Directorships, 1839-1919" Belknap Press. Cam. Mass.
Keynes. G. 1960. "A Bibliography of Dr. Robert Hooke." Oxford. Clarendon Press.
Kopal. Z. "The Dictionary of Scientific Biography." Charles Schribener and Sons. New York.
Lambert, C. "An Aristocrat's Killing." Harvard magazine. July/Aug 2003
Landau & Lifshitz. "Quantum Electrodynamics/" Butterworth Heinmann. Oxford.
Leggat, R. 2000. "A History of Photography." (www.rleggat.com/photohistory/)
Lerner. E.J. "The Big Bang Never Happened" Vintage Books. New York.
Livio. M. "The Accelerating Universe." John Wiley & Son. New York.
Longair. M.S. "High Energy Astrophysics." Volumes 1 & 2.Cambridge University Press. Cambridge.
Lovi. G. 1985. "The Distance Dilemma." Sky & Telescope 69. 45-46.
McCready, D. "The Moon Illusion Problem." Paper presented to colloquia at:
The University of Chicago (1964); Marquette University (1968); Lawrence University; The University of Wisconsin-Whitewater. http://facstaff.uww.edu/mccreadd/bibliography.html
Mitchell. W.C. "Bye Bye Big Bang." Cosmic Sense books. Nevada.
"Murder at Harvard." http://www.news.harvard.edu/gazette/2002/10.03/23-murder.html

NASA & Mars orbital loss.
http://www.cnn.com/TECH/space/9909/30/mars.metric/
Nichols. R. 1999. "Robert Hooke and the Royal Society." Sussex UK. The Book Guild.
Park. D. 1997. "The Fire Within the Eye; A Historical Essay on the Nature and Meaning of Light." Princeton. Princeton University Press.
Paterson. R. 1957. "Robert Hooke." Sky & Telescope 16. 179-180.
"Robert Hooke: Victim of Genius" BBC Documentary.
http://www.bbc.co.uk/bbcfour/documentaries/features/hooke.shtml
Russell. J.B. "The Myth of the Flat earth." American Affiliation Conference 1997. Westmont College.
http:/www.id.ucsb.edu/fscf/library/RUSSELL/FlatEarth.html
Schindler, K.
http://www.lowell.edu/online_newsletter/spring_03/slipher.html
Sigismondi. C. "Mira Ceti and the Star of Bethlehem."http://www.quodlibet.net/sigismondi-mira.shtml
Simons. M. "The History of Mount Wilson Observatory." Mount Wilson Observatory Association.
http://www.mtwilson.edu/his/art/g1a4.htm
Starlore: Origins of the Constellations.
http://www.stargazers.net.au/constellorigins.htm
Stone. B. "Antinous. Last of the roman Gods.'
http://www.users.bigpond.net.au/bstone/antinous.htm
Van Helden. 1985. "Measuring the Universe; Cosmic Dimensions from Aristarchus to Halley." Chicago. University of Chicago Press.
"Was an Island Hooke Discovery Suppressed by Academia?" Isle of White History Centre.
http://www.freespace.virgin.net/iw.history/archive/newsjan4.htm
Whitney. C. 1971. "The Discovery of the Galaxy." New York. Alfred A. Knopf.
White. M. "Isaac Newton. The Last Sorcerer." Forth Estate London.
"Women in Astronomy." Astronomical Society of the Pacific.

291
eyy + collapse

296 in tune
look Back different

Made in the USA
Lexington, KY
02 March 2012